THE
WISDOM
OF THE
SAINTS

THE
WISDOM
OF THE
SAINTS

AN ANTHOLOGY

Jill Haak Adels

NEW YORK   OXFORD
OXFORD UNIVERSITY PRESS
1987

Oxford University Press

Oxford   New York   Toronto
Delhi   Bombay   Calcutta   Madras   Karachi
Petaling Jaya   Singapore   Hong Kong   Tokyo
Nairobi   Dar es Salaam   Cape Town
Melbourne   Auckland

and associated companies in
Beirut   Berlin   Ibadan   Nicosia

Copyright © 1987 by Jill Haak Adels

Published by Oxford University Press, Inc.,
200 Madison Avenue, New York, New York 10016

Oxford is a registered trademark of Oxford University Press

Library of Congress Cataloging-in-Publication Data
The wisdom of the saints.
1. Meditations.   I. Adels, Jill Haak.
BX 2178.W57 1987   242   86-12626
ISBN 0-19-504152-6

2 4 6 8 10 9 7 5 3 1
Printed in the United States of America
on acid-free paper

I have not wished to enrich the edition with any references, as some have desired me to do, because the learned do not need such things, and the others do not bother about them.   ST. FRANCIS DE SALES

PREFACE

This is a collection of the the saints' own words, from the time of the early fathers of the church to our own day. The selections are brief and are arranged thematically. Their literary quality ranges from the great polish and sophistication of the writings of St. Francis de Sales or St. Thomas More to the exclamations of the martyr on the scafford. Their tone varies from the intimacy of letters written for one pair of loving eyes to see to sermons preached to thousands with apostolic eloquence. Lists of personal resolutions, diaries, half-legendary remarks from the early wonder-working saints of Ireland or the northern forests, naive remarks of saintly children—later remembered and assigned a deep significance—all are here.

One of the vital functions of a living church is to lift us out of the here and now into a context of eternity. On a human level, the image of the church as a timeless communion of all who have loved God, whatever their century, sex, social class, or cultural circumstances, is both comforting and inspiring. As a human family that transcends time, the saints tell us that our struggles are of immense significance and that they are understood. Understanding the human struggles and joys described in these pages adds depth and significance to the challenges we face. The courage and wisdom of the saints should heighten ours.

Not all the saints who speak from these pages have been canonized by the church, which, of course is a very different thing from saying that they are not enjoying the benefits of sainthood. Some of the most eloquent, like Theophane Venard or Henry Suso, are "merely" beatified. I have been unable to resist including a number of sayings from the anonymous author of *The Cloud of Unknowing* because of the book's lucid charm and the way it makes tremendous questions simple and appealing. There is also a specific appeal in the anonymity of the author—the "unknown saint," like the unknown soldier, could stand for many. The old peasant of Ars who makes his simple comment on the secret of contemplative prayer similarly may stand for the legion of saintly people who have lived lives of great obscurity and have never found a chronicler. Some of them are certainly living, or have lived, in one's own town and family.

Some canonized saints have been literary geniuses; Augustine, Bernard, and Teresa of Avila may stand for a distinguished company. Others are almost unreadable, at least by me; Saints Margaret Mary Alacoque and Gertrude the Great spring instantly to mind. Members of a third group have a utilitarian or nonexistent literary style but fascinate and delight by strength of personality, the charm of simplicity, or the compelling force of what they have to say. There is an authoritative, authentic voice of sainthood which can be recognized across the centuries, across cultural and temporal boundaries.

The very idea of saints, except for those who seem particularly relevant for our time, like Maximilian Kolbe who died at Auschwitz, is currently unfashionable. Francis of Assisi, for his "nature mysticism," John of the Cross for his poetry, and Teresa of Avila because she was a Great Woman are among the very few who are accorded some status by the general culture today. As a "pre-saint," Mother Teresa of Calcutta is cast very much in a traditional mold, but even the popular religious press seems unaware that she has heroic precursors by the thousand. This situation is bound to change, as all such cycles do, but meanwhile a generation is missing an experience that is of great value.

On a purely human level there are many treasures to be found in the literature of saintliness. As Phyllis McGinley put it in one of the few "saint books" that has maintained its popular appeal:

> They lost their tempers, got hungry, scolded God, were egotistical or testy or impatient in their turns, made mistakes and regretted them. Still they went on doggedly blundering toward Heaven. And they won sanctity partly by willing to be saints, not only because they encountered no temptation to be less.

François Mauriac adds:

> Hagiography exalts the saints as if they no longer shared the human condition. To tell the truth, to the extent that they are lovers, that they are literally mad with love, the saints, like the rest of us, remain subject to strange and terrible mistakes.

If old-fashioned hagiographical exaltation is partly responsible for our current lack of interest in the saints, then our current neglect may be a useful reaction. Later, perhaps, we can make a fresh start toward interpreting the news they send us from the frontiers of spiritual experience. Meanwhile, this collection may serve as one reminder of the treasures that wait for us. I have chosen to let the saints speak for themselves; their voices blend here in affirmation and contradiction, paradox or harmony, praising earthly and spiritual joys, the exquisite agonies of purgation, and the bliss of divine love.

South Hadley, Massachusetts   J.H.A.
April 1986

CONTENTS

THE
WISDOM
OF THE
SAINTS

SAINTS

The name of "saint" is used in the New Testament as a synonym for "Christian" but soon was reserved for martyrs who had died in defense of the faith or for those of exceptional holiness. The Roman Martyrology lists more than five thousand "canonized" or officially recognized saints who are members of the Church Triumphant, presumed to be in Heaven. There are, of course, many more, unknown saints. Those granted the title have led lives of conspicuous, heroic virtue and, in recent centuries, have gone through a torturous bureaucratic process requiring indisputable proof of their excellence, which may take centuries for its completion.

The titles of "Venerable" and "Blessed" are signs that a candidate has passed through preliminary stages and is eligible for the ultimate accolade. Far from abolishing the saints, as many people vaguely suppose, Vatican II accorded due respect to the stately process of canonization and simply dismissed as "not proven" on grounds of insufficient evidence the cause of legendary figures like St. Christopher. The saints who remain are all real people whose lives are held to be a witness and example to the Christian faithful.

The saints here introduce each other and the topic of sainthood. They are well qualified to do so; many have written lives of other saints, or have known other saints well as family members, friends, disciples and mentors.

Many have also described themselves as purposefully striving to become saints and have predicted their own apotheosis. This seems almost shockingly bold, scarcely compatible with humility, but it merely takes seriously Christ's commandment to "be perfect just as your heavenly father is perfect" (Matthew 5:48).

Saints discuss each other with admiration and affection. They tell inspiring anecdotes and celebrate the reality of the communion of saints by giving each other messages and tasks to be carried out from Heaven.

A third theme could be called the aesthetics and etiquette of sainthood. We are warned not to make saints and their achievements meaningless by portraying them as faultless superhumans or, conversely, by awarding the title lightly.

———

The saints have no need of honor from us; neither does our devotion add the slightest thing to what is theirs. Clearly, if we venerate their memory, it serves us, not them. But I tell you, when I think of them, I feel myself inflamed by tremendous yearning.   BERNARD OF CLAIRVAUX

This daring ambition of aspiring to great sanctity has never left me. I don't rely on my own merits, because I haven't any; I put all my confidence in him who is virtue, who is holiness itself.   THÉRÈSE DE LISIEUX

I would be the happiest of men if I could become a saint soon and a big one.   JOSEPH CAFASSO

I am fortunate to be given this opportunity to become a saint. I want to become a saint. If I bungle in this life, I will spoil things forever. If things turn out well, and I am equal to the task, I will achieve happiness and become a saint.   BL. MARY FORTUNATA VITI

I greatly desire to become a saint that I may be able to make saints and thus procure the glory of God.   PETER JULIAN EYMARD

I see that my chief obstacle to holiness is pride. I *will* overcome it!   VEN. ANDREW BELTRAMI

A time will come when you will call me a saint, and you will go on a pilgrimage to my tomb with the staff and wallet of the pilgrim.   MARGARET OF CORTONA

If you embrace all things in life as coming from the hands of God, and even embrace death to fulfill His holy will, assuredly you will die a saint.   ALPHONSUS LIGUORI

The Saints were so completely dead to themselves that they cared very little whether others agreed with them or not.   JOHN VIANNEY

The whole science of the Saints consists in knowing and following the will of God.   ISIDORE OF SEVILLE

The history of the Church teaches us that the greatest saints are those who professed the greatest devotion to Mary.   JOHN BOSCO

What, then, does God look upon with pleasure and delight? What is He asking the Angels about, and even the devils? It is about the man who is fighting for Him against riches, against the world, hell, and himself, the man who is cheerfully carrying his cross.

LOUIS-MARIE GRIGNION DE MONTFORT

There is no harm to the saints if their faults are shown as well as their virtues. But great harm is done to everybody by the hagiographers who slur over the faults, be it for the purpose of honoring the saints . . . or through fear of diminishing our reverence for their holiness.

FRANCIS DE SALES

Sanctity does not consist in being odd, but it does consist in being rare.

FRANCIS DE SALES

In all those things which do not come under the obligation of faith, the saints were at liberty to hold divergent views, just as we ourselves are.

THOMAS AQUINAS

Those from whom I receive the greatest consolations and encouragement are those whom I know to be dwelling in Paradise.

TERESA OF AVILA

The inimitable things in the saints often do us the most good.

TERESA OF AVILA

Theophane Venard was such a *little* saint. His life contains nothing out of the ordinary. He loved our Lady Immaculate and he loved his family as well. So do I. I simply cannot understand those saints who didn't. As a farewell keepsake for you I have copied some passages from his last letters to his relatives; they are my very thoughts. My soul is like his soul.

THÉRÈSE DE LISIEUX

I am sending you the *Life of St. Catherine of Siena*, who is my Mistress and Directress. She moves and touches me so deeply that each time I read her life, I have to hold the book in one hand and a handkerchief in the other to staunch the tears that it makes me shed continually.

ANTHONY MARY CLARET

In beautiful things St. Francis saw Beauty itself, and through His vestiges imprinted on creation he followed his Beloved everywhere, making from all things a ladder by which he could climb up and embrace Him who is utterly desirable.

BONAVENTURE

His call came before the first hour, when he was only a child. He used to tell me that he considered his seventh year to have been the year of his conversion. Then again, he never suffered from fleshly allurements, even in his thoughts, and he is the only one I know who was so singularly blessed. His third privilege was to be free of all distractions in his prayers, and how great a privilege that was, we who try to pray know best.

ROBERT BELLARMINE, *on Aloysius Gonzaga*

He was Firm by name and grace made him Firmer.

AUGUSTINE OF HIPPO, *speaking of St. Firmus*

Saint John appears as the boundary between the two testaments; . . . As representative of the past, he is born of aged parents; as a herald of the new era, he is declared to be a prophet while still in his mother's womb. For while yet unborn he leapt in his mother's womb at the arrival of the blessed Mary.   AUGUSTINE OF HIPPO

Can't you send us a whole squadron of Francis Xaviers for this China of ours which stands in such need of them?   BL. JOHN GABRIEL PERBOYLE

Avoid all comparison of living men, however great their merit, with the saints; for example, avoid saying: 'Such a one is more learned than St. Augustine; this is another St. Francis; this one is as zealous, as eloquent, as St. Paul. . .   IGNATIUS OF LOYOLA

The soul doth oftentimes more readily pray for assistance unto the saints rather than unto God, not daring because of its unworthiness to pray unto God Himself, albeit it imploreth help from the Holy Virgin and all the other saints.   BL. ANGELA OF FOLIGNO

Often make acts of love of our Lady, the saints, and the holy angels. Make friends with them. Talk with them frequently, using words of praise and tenderness. When you have gained familiar access to the citizens of the heavenly Jerusalem above, you will grieve far less at bidding farewell to those of the mean city here below.   FRANCIS DE SALES

During her last agony I saw a tear shining like a diamond on her eyelashes, the last she would ever shed. It didn't fall; I could see it still shining there when she lay in choir, and as nobody had removed it, I went in quietly that evening with a piece of fine linen, and took this for my relic—the last tear of a saint.   THÉRÈSE DE LISIEUX

We had each received from Mère Marthe a little scrap of the vestment of St. Vincent. I cut mine in two and swallowed half before going to sleep, persuaded that St. Vincent would obtain for me the grace of seeing the Blessed Virgin.   CATHERINE LABOURÉ

Think how many saints there are in heaven who see their fathers, mothers, brothers, and other relatives in the damnation of hell, which is the misfortune of misfortunes and the height of all woes; and notwithstanding they adore, they love, they bless with joy and happiness that most just will because they see that such is the decree of divine justice concerning these relatives. JEAN EUDES

Open, O Lord, the Book of Life in which are written the deeds of Thy saints; all the deeds told in that book I long to have accomplished for Thee. I would be a Martyr, a Doctor of the Church. I should like to accomplish the most heroic deeds—the spirit of the Crusader burns in me, and I long to die on the battlefield in defense of Holy Church. I would be a Missionary. I would be flayed like St. Bartholomew, plunged into boiling oil like St. John, or, like St. Ignatius of Antioch, I would be ground by the teeth of wild beasts into bread worthy of God. With St. Agnes and St. Cecilia I would offer my neck to the sword of the executioner, and like St. Joan of Arc I would murmur the name of Jesus at the stake. THÉRÈSE DE LISIEUX

Anthony did, in fact, heal without issuing commands, but by praying and calling on the name of Christ, to make it clear to all that it was not he who did this but the Lord, bringing his benevolence to effect through Anthony and curing those who were afflicted. ATHANASIUS, *The Life of Anthony*

The greater the charity of the saints, in their heavenly home, the more they intercede for those who are still on their journey and the more they can help them by their prayers; the more they are united with God, the more effective those prayers are. This is in accordance with divine order, which makes higher things react upon lower things, like the brightness of the sun filling the atmosphere. THOMAS AQUINAS

We must never lose sight of the fact that we are either Saints or outcasts, that we must love for Heaven or for Hell; there is no middle path in this.
JOHN VIANNEY

GOD

Again and again the saints say that experience is the only road to God. It can be emotional experience, as in fear or love, or the experience of faith, which then leads to understanding. Of all forms of experience, mystical states are the most compelling and convincing.

Non-mystics can look at the works of God and draw conclusions from the sense of wonder they evoke. All emotions that God inspires in us have immense value, including "servile fear."

"I am who am" (Exodus 3:14). God's infinite meaningfulness guarantees that all creation is laden with meaning and value. The certainty that nothing is without symbolism pervades medieval thought, which sought purpose and allegory in every connecting link of the great chain of being. It is the farthest attitude imaginable from the modern sense of absurdity and accident, of mechanical cause and effect.

The Trinity exists, because, as Ruysbroeck says, only God is capable of experiencing the degree of love that He deserves. To the Holy Spirit are attributed all the works of the Trinity that pertain to the sanctification of the human race.

He who desires nothing but God is rich and happy.   ALPHONSUS LIGUORI

This is the Creator: in respect of his love, our Father, in respect of his power, our Lord, in respect of his wisdom, our Maker and Designer.
                                                                    IRENAEUS

Describe him as you will: good, fair Lord, sweet, merciful, righteous, wise, all-knowing, strong one, almighty; as knowledge, wisdom, might, strength, love, or charity, and you will find them all hidden and contained in this little word *is*.   ANONYMOUS, *The Book of Privy Counseling*

Though the persons of the blessed Trinity be all alike in their attributes, it was their love that was the most shown to me and that it is closest to us all.   JULIAN OF NORWICH

God is never loved according to his worth by any creatures. And to the enlightened reason this is a great delight and satisfaction: that its God and its Beloved is so high and so rich that He transcends all created powers, and can be loved according to His merits by none save Himself.

BL. JAN VAN RUYSBROECK

When God said, "Let there be light," he did not speak in order that some subordinate might hear, understand what the speaker wanted, and go and perform the task. This is what happens in human affairs. But the Word of God is creator and maker, and he *is* the Father's will.   ATHANASIUS

God willed that His Son suffer inconceivably cruel and horrible torments; not only that He suffer them but that He die the most shameful and atrocious death of all possible deaths! Oh, how severe is the will of a Father regarding His Son! How strange and terrible it is!   JEAN EUDES

The Spirit works, the Son fulfills his ministry, and the Father approves; and man is thus brought to full salvation.   IRENAEUS

Holy Spirit, Spirit of truth, you are the reward of the saints, the comforter of souls, light in the darkness, riches to the poor, treasure to lovers, food for the hungry, comfort to those who are wandering; to sum up, you are the one in whom all treasures are contained.   MARY MAGDELENE DEI PAZZI

In God alone is there primordial and true delight, and in all our delights it is this delight that we are seeking.   BONAVENTURE

God has his reasons for everything that comes to pass. If God were not God, if he had never revealed himself, we might not know what to make of it.   BL. PLACID RICCARDI

God gave the elephant the largest bodily proportions so that it could carry the heaviest burdens and even howdahs filled with men. He made the horse a little smaller, just large enough in fact for carrying one rider. The birds He made small enough to hang nests in the branches of trees. And the bees and ants He made very tiny to fit them for concealment in anthills, beehives, and holes in the ground.   ROBERT BELLARMINE

I may turn myself hither, I may turn myself thither, in me is nothing that can displease; in me is everything that can delight the inmost wishes of the heart and all the desires of the soul.   BL. HENRY SUSO

God has so fixed the number of parts that no additions or subtractions may be made. Thus He has given to a man two eyes, two ears, two hands, two feet, one nose, one mouth, one chest, and one head: The result is something very beautiful and well-proportioned. But upset that order, apportion to any man one eye, two noses, one ear, two mouths, one hand and one foot, two trunks and two heads: you would find nothing more disgusting or more useless.   ROBERT BELLARMINE

Although our view of the sublimest things is limited and weak, it is most pleasant to be able to catch but a glimpse of them.   THOMAS AQUINAS

It is sometimes necessary to speak to this mighty All, and be ready for our nothing to face an encounter with something.   FRANCIS DE SALES

Imagine a man in whom the tumult of the flesh goes silent, in whom the images of earth, of water, of air, and of the skies cease to resound. His soul turns quiet and, self-reflecting no longer, it transcends itself. Dreams and visions end. So too does all speech and every gesture, everything in fact which comes to be only to pass away. All these things cry out: "We did not make ourselves. It is the Eternal One who made us."

AUGUSTINE OF HIPPO

It is true that the voice of God, having once fully penetrated the heart, becomes strong as the tempest and loud as the thunder; but before reaching the heart it is as weak as a light breath which scarcely agitates the air. It shrinks from noise, and is silent amid agitation.   IGNATIUS OF LOYOLA

He alone is God who can never be sought in vain; not even when he cannot be found.   BERNARD OF CLAIRVAUX

We are speaking of God, what marvel if thou do not comprehend? For if thou comprehend, He is not God.   AUGUSTINE OF HIPPO

When I do reflect within myself, I do sometimes perceive most clearly that those persons who do best know God (Who is infinite and unspeakable) are those who do the least presume to speak of Him, considering that all which they do say of Him, or possibly can say, is as nothing compared with what He truly is.   BL. ANGELA OF FOLIGNO

How could there be any other Totality beyond him, or another Principle or Power or another God? For God who is the totality of all these must include all things in his infinite being.   IRENAEUS

If you contemplate God with the eyes of faith, you will see Him just as He is, and, in a certain manner, face to face.   JEAN EUDES

With God nothing is empty of meaning, nothing without symbolism.
IRENAEUS

Although it is very praiseworthy and useful to serve God through the motive of pure charity, yet we must also recommend the fear of God; and not only filial fear, but servile fear, which is very useful and often even necessary to raise man from sin. Once risen from the state and free from mortal sin, we may then speak of that filial fear which is truly worthy of God and which gives and preserves the union of pure love. IGNATIUS OF LOYOLA

Let us love God, but with the strength of our arms, in the sweat of our brow. VINCENT DE PAUL

Why, then has the fool said in his heart, there is no God, when it is so evident, to a rational mind, that thou dost exist in the highest degree of all? Why, except that he is dull and a fool? ANSELM OF CANTERBURY

The Father of all has no name given him since he is unbegotten. For a being who has a name imposed on him has an elder to give him that name. "Father" and "God," "Creator," "Lord," "Master," are not names but appellations derived from his benefits and works. JUSTIN MARTYR

When the heart is pure it cannot help loving, because it has discovered the source of love which is God. JOHN VIANNEY

THE INCARNATION

The Incarnation is the central doctrine of Christianity; "The Word was made flesh and dwelt among us" (John 1:14). God assumed our flesh, body, and soul in order to redeem us. The obligations imposed on humans by God's condescension are staggering. The elevation and healing of human nature are also implied. That which was made of earth can now pass, as Athanasius tells us, "through the gates of heaven."

Who could find it hard to persevere at the sight of a God who never commands us to do anything which he has not practiced himself?

JOHN VIANNEY

Christ did not pass through the Virgin as through a channel, but truly took flesh and was truly fed with milk from her. He truly ate as we eat and drank as we drink. For if the incarnation was a figment then our salvation was a figment.

CYRIL OF JERUSALEM

Moses built the house, yet did not complete it and went away. Then God raised up the choir of prophets by his spirit, and they built on the foundation of Moses, but could not complete it and also went away. All of them, through the spirit, saw that the wound was incurable, and that none of the creatures was able to heal it, except the Only Begotten, the very Mind of the Father, and His very Image.

ANTHONY OF EGYPT

Invisible in his own nature he became visible in ours. Beyond our grasp, he chose to come within our grasp. Existing before time began, he began to exist at a moment in time. Incapable of suffering as God, he did not refuse to be a man, capable of suffering. Immortal, he chose to be subject to the laws of death.

LEO THE GREAT

If you could see the sweet embrace of the Virgin and the woman who had been sterile and hear the greeting in which the tiny servant recognized his

Lord, the herald his Judge, and the voice his Word, then I am sure you would sing in sweet tones with the Blessed Virgin that sacred hymn: *My soul magnifies the Lord*. . . .   BONAVENTURE

He did submit himself unto the elements, unto cold and heat, hunger and thirst, and other insensible creatures, concealing His power and despoiling Himself thereof in the likeness of man, in order that He might teach us weak and wretched mortals with what patience we ought to bear tribulation.

BL. ANGELA OF FOLIGNO

He undertook to help the descendants of Abraham, fashioning a body for himself from a woman and sharing our flesh and blood, to enable us to see in him not only God, but also, by reason of this union, a man like ourselves.   CYRIL OF ALEXANDRIA

Christ is born that by his birth he might restore our nature. He became a child, was fed, and grew that he might inaugurate the one perfect age to remain forever as he created it. He supports man that man might no longer fall. And the creature he had formed of earth he now makes heavenly.

PETER CHRYSOLOGUS

Man's body has acquired something great through its communion and union with the Word. From being mortal it has been made immortal; though it was a living body it has become a spiritual one; though it was made from the earth it has passed through the gates of Heaven.   ATHANASIUS

Through Christ we see as in a mirror the spotless and excellent face of God.

CLEMENT OF ROME

The Word was not degraded by receiving a body, so that he should seek to "receive" God's gift. Rather he deified what he put on; and, more than that, he bestowed this gift upon the race of men.   ATHANASIUS

To pay the debt of our sinful state, a nature that is incapable of suffering was joined to one that could suffer. Thus, in keeping with the healing that we needed, one and the same mediator between God and men, the man Jesus Christ, was able to die in one nature, and unable to die in the other.

LEO THE GREAT

CHRIST

The Second Person of the Holy Trinity, the eternal Word of God, who was made flesh to redeem the world from sin and death by his sacrifice on the cross, Christ is the image of the invisible God and also the perfect man. "For God so loved the world that he gave his only-begotten son, that those who believe in him should not perish, but have everlasting life" (John 4:14).

The name of Christ means "the annointed one, the Messiah, the prophesied king," identifying him with the Promised One of Israel come to deliver his people from their godless enemies and to reign over them in faithfulness and righteousness. He was executed in Jerusalem by the Roman rulers of Palestine in the first century A.D., as a pretender to the title of "the King of the Jews." Other titles given him in the New Testament include the Lamb of God, the Good Shepherd, Immanuel, the Chosen One, the Morning Star, the True Vine, the Living Bread, the Son of God, the Son of Man, Son of Mary, Carpenter, and Alpha and Omega.

There is one physician, fleshly and spiritual, begotten and unbegotten, God in man, true life in death, both of Mary and of God, first passible then impassible, Jesus Christ our Lord.   IGNATIUS OF ANTIOCH

When we speak about wisdom, we are speaking of Christ. When we speak about virtue, we are speaking of Christ. When we speak about justice, we are speaking of Christ. When we speak about peace, we are speaking of Christ. When we speak about truth and life and redemption, we are speaking of Christ.   AMBROSE

As no darkness can be seen by anyone surrounded by light, so no trivialities can capture the attention of anyone who has his eyes on Christ. The man who keeps his eyes upon the head and origin of the whole universe has them on virtue in all its perfection; he has them on truth, on justice, on immortality and on everything else that is good, for Christ is goodness itself. GREGORY OF NYSSA

The almighty and most holy Word of the Father pervades the whole of reality, everywhere unfolding his power and shining on all things visible and invisible. He sustains it all and binds it together in himself. He leaves nothing devoid of his power but gives life and keeps it in being throughout all of creation and in each individual creature. ATHANASIUS

Take thought now, redeemed man, and consider how great and worthy is he who hangs on the cross for you. His death brings the dead to life, but at his passing heaven and earth are plunged into mourning and hard rocks are split asunder. BONAVENTURE

Out of love the Lord took us to himself; because he loved us and it was God's will, our Lord Jesus Christ gave his life's blood for us—he gave his body for our body, his soul for our soul. CLEMENT OF ROME

Where have your love, your mercy, your compassion shone out more luminously than in your wounds, sweet gentle Lord of mercy? More mercy than this no one has than that he lay down his life for those who are doomed to death. BERNARD OF CLAIRVAUX

Conceal yourselves in Jesus crucified, and hope for nothing except that all men be thoroughly converted to his will. PAUL OF THE CROSS

Do you want to honor Christ's body? Then do not scorn him in his nakedness, nor honor him here in the church with silken garments while neglecting him outside where he is cold and naked. JOHN CHRYSOSTOM

Christ hungers now, my brethren; it is he who deigns to hunger and thirst in the persons of the poor. And what he will return in heaven tomorrow is what he receives here on earth today. CAESARIUS OF ARLES

I desire and choose poverty with Christ poor, rather than riches; insults with Christ loaded with them, rather than honors; I desire to be accounted as worthless and a fool for Christ rather than to be esteemed as wise and prudent in this world. So Christ was treated before me.

IGNATIUS OF LOYOLA

Since the soul joined to a frail body can never unite itself so intimately to the pure and sovereign good as the greatness and sublimity of such an alliance demand, it must choose certain holy and divine images which are able to draw it away from itself and raise it to God. Among such images the first is the image and example of Jesus Christ, God and Man, the maker of all the saints, in whom is found life itself, the reward and happiness of the soul.   BL. HENRY SUSO

Meekness was the method that Jesus used with the apostles. He put up with their ignorance and roughness and even their infidelity. He treated sinners with a kindness and affection that caused some to be shocked, others to be scandalized, and still others to gain hope in God's mercy. Thus, he bade us to be gentle and humble of heart.   JOHN BOSCO

A young village girl told me, When I am about to talk to anyone, I picture to myself Jesus Christ and how gracious and friendly he was to everyone."
                                                                                    JOHN VIANNEY

I will imagine that my soul and body are like the two hands of a compass, and that my soul, like the stationary hand, is fixed in Jesus, who is my center, and that my body, like the moving hand, is describing a circle of assignments and obligations.   ANTHONY MARY CLARET

I no longer wish to find happiness in myself or in created and perishable things, but in Jesus my savior. He is my All, and I desire to belong wholly to Him. It is the most extreme folly and delusion to look elsewhere for any true happiness. Let us, then, vehemently and courageously renounce all other things and seek only Him.   JEAN EUDES

If, then, you are looking for the way by which you should go, take Christ, because he himself is the way.   THOMAS AQUINAS

THE BLESSED VIRGIN

Mary, who is virtually the only venerated feminine image of the divine to be found in the monotheistic religions, ought to be of particular interest in our era as we struggle to modify or enrich patriarchal traditions.

Within the tradition of the Church itself, it is striking to realize how early, how complete, and how warmly loving is the picture of Mary and her role. The earliest saints quoted here are as ready to see Mary as the queen of heaven or the "heart of the Church" as are the medieval and modern saints.

He who wrote on the tablets of stone without iron made Mary with child of the Holy Ghost; and He who produced bread in the desert without ploughing impregnated the Virgin without corruption; and He who made the rod to bud without rain made the daughter of David bring forth without seed.

AUGUSTINE OF HIPPO

The Divine Spirit, the love itself of the Father and the Son, came corporally into Mary, and enriching her with graces above all creatures, reposed in her and made her His Spouse, the Queen of Heaven and earth.

ANSELM OF CANTERBURY

She had the noblest son that ever woman was mother of, not only above the condition of men, but above the glory of angels; being her son only, without temporal father, and thereby doubling the love of both parents in her breast.

ROBERT SOUTHWELL

The Blessed Virgin, by becoming the Mother of God, received a kind of infinite dignity because God is infinite; this dignity therefore is such a reality that a better is not possible, just as nothing can be better than God.

THOMAS AQUINAS

That anyone could doubt the right of the holy Virgin to be called the Mother of God fills me with astonishment. Surely she must be the mother of God if our Lord Jesus Christ is God, and she gave birth to him! Our Lord's disciples may not have used those exact words, but they delivered to us the belief those words enshrine, and this has also been taught to us by the holy fathers.   CYRIL OF ALEXANDRIA

Mary is the heart of the church. This is why all works of charity spring from her. It is well known that the heart has two movements: systole and diastole. Thus Mary is always performing these two movements: absorbing grace from her Most Holy Son, and pouring it forth on sinners.
                                ANTHONY MARY CLARET

The Immaculate alone has from God the promise of victory over Satan. She seeks souls that will consecrate themselves entirely to her, that will become in her hands forceful instruments for the defeat of Satan and the spread of God's kingdom.   MAXIMILIAN KOLBE

Men do not fear a powerful hostile army as the powers of hell fear the name and protection of Mary.   BONAVENTURE

At the mention of this name the angels rejoice and the devils tremble; through this invocation sinners obtain grace and pardon.   PETER CANISIUS

As flies are driven away by a great fire, so were the evil spirits driven away by her ardent love for God.   BERNARDINO OF SIENA

St. Anselm tells us that salvation is occasionally more obtained by calling on the name of Mary than by invoking that of Jesus. This is not because He is not the source and Lord of all graces, but because, when we have recourse to the Mother, and she prays for us, her prayers—the prayers of a mother—are more irresistible than our own.   ALPHONSUS LIGUORI

There is no more excellent way to obtain graces from God than to seek them through Mary, because her Divine Son cannot refuse her anything.
                                PHILIP NERI

Her Son esteems her prayers so greatly, and is so eager to please her that when she prays it seems as if she rather commands, and is rather a queen than a handmaid.   PETER DAMIAN

She is so beautiful that to see her again one would be willing to die.
                                BERNADETTE SOUBIROUS

The beauty that I saw in Our Lady was extraordinary, although I didn't make out any particular details except the form of her face in general and that her garment was of the most brilliant white, not dazzling, but soft. Our Lady seemed to me to be a very young girl.　　　TERESA OF AVILA

I saw her ghostly, in bodily likeness: a simple maid and meek, young of age, and little waxen above a child, in the stature that she was when she conceived.　　　BL. JULIAN OF NORWICH

Her eyes were turned toward heaven, she stood erect upon a large white sphere, her feet were set upon a serpent, greenish in color but touched with yellow spots. At the level of her breast she held a little golden ball surmounted by a Cross, and this she was offering to God.

CATHERINE LABOURÉ

On Matins of the feast of the Annunciation I saw the heart of the Virgin Mother so bathed by the rivers of grace flowing out of the Blessed Trinity that I understood the privilege Mary has of being the *most powerful* after God the Father, the *wisest* after God the Son, and the *most benign* after God the Holy Spirit.　　　GERTRUDE THE GREAT

Suddenly my soul was uplifted to behold and contemplate the Blessed Virgin Mary in Glory; and beholding a woman placed in such nobility, glory, and dignity as was she, I was filled with marvelous delight, and the sight did produce in me most immense joyfulness. The glorious Virgin was praying for the human race.　　　BL. ANGELA OF FOLIGNO

If you invoke the Blessed Virgin when you are tempted, she will come at once to your help, and Satan will leave you.　　　JOHN VIANNEY

If you are in danger, she will hasten to free you. If you are troubled, she will console you. If you are sick, she will bring you relief. If you are in need, she will help you. She does not look to see what kind of a person you have been. She simply comes to a heart that wants to love her.

GABRIEL POSSENTI

People may well say that Don Bosco sees everything clearly and that the Blessed Virgin leads him step by step. At every moment, in every circumstance, she reveals herself. She has been protecting me from every danger, she shows me the task to be accomplished, and she always helps me to carry it out.　　　JOHN BOSCO

When the holy spirit finds that a soul has drawn close to Mary, His dear and inseparable spouse, He quickens His activity of forming Jesus Christ in it.　　　LOUIS-MARIE GRIGNON DE MONTFORT

Mother dear, lend me your heart. I look for it each day to pour my troubles into.　　　GEMMA GALGANI

My mother is very strange; if I bring her flowers, she says she does not want them; if I bring her cherries, she will not take them, and if I then ask her what she desires, she replies: "I desire thy heart, for I live on hearts."   JOSEPH OF COPERTINO

As often as the sweet name of Mary comes to your lips, you ought to represent to yourself a masterpiece of God's power so perfect and sublime that even the arm of the Almighty could not produce anything more perfect in the shape of a pure creature.   LEONARD OF PORT MAURICE

Divine love so inflamed her that nothing earthly could enter her affections; she was always burning with this heavenly flame and, so to say, inebriated with it.   SOPHRONIUS

Divine love so penetrated and filled the soul of Mary that no part of her was left untouched, so that she loved with her whole heart, with her whole soul, and her whole strength, and was full of grace.
                                                BERNARD OF CLAIRVAUX

The Holy Ghost heated, inflamed, and melted Mary with love, as fire does iron; so that the flame of this Holy Spirit was seen and nothing was felt but the fire of the love of God.   ILDEPHONSUS

The bush seen by Moses, which burnt without being consumed, was a real symbol of Mary's heart.   THOMAS OF VILLANOVA

Behold the power of the Virgin Mother: she wounded and took captive the heart of God.   BERNARDINO OF SIENA

With reason is Mary called the Virgin of virgins: for she, without the counsel or example of others, was the first who offered her virginity to God.   ALBERT THE GREAT

The Saints and Angels are but compared to stars, and the first of them to the fairest of the stars; but she is fair as the moon, as easy to be chosen and discerned from all the Saints as the sun from the stars.   FRANCIS DE SALES

Your name, O Mary, is a precious ointment, which breathes forth the odor of Divine grace. Let this ointment of salvation enter the inmost recesses of our souls.   AMBROSE

O name of Mary! Joy in the heart, honey in the mouth, melody to the ear of her devout clients!   ANTHONY OF PADUA

You, O Mary, are the young plant that produced the fragrant flower of the Word, only begotten of God, because you were the fertile land that was sown with this Word.   CATHERINE OF SIENA

Hail, God's palace; hail Tabernacle of the Most High; Hail, House of God; Hail, his Holy Vestments; Hail, Handmaid of God.   FRANCIS OF ASSISI

Glory of virgins, the joy of mothers, the support of the faithful, the diadem of the Church, the model of the true Faith, the seat of piety, the dwelling place of the Holy Trinity.   PROCLUS OF CONSTANTINOPLE

Through thee idolatrous creatures have known incarnate truth, the faithful have received baptism, churches have been erected in all parts of the earth. By thine assistance the Gentiles have been brought to repentance. And finally, through thee, the only Son of God, source of all light, has shone upon the eyes of the blind, who were sitting in the shadow of death.
                    CYRIL OF JERUSALEM

He who is devout to the Virgin Mother will certainly never be lost.
                    IGNATIUS OF ANTIOCH

Cling to His most sweet Mother who carried a son Whom the heavens could not contain; and yet she carried Him in the little enclosure of her holy womb and held him on her virginal lap.   CLARE OF ASSISI

To serve Mary and to be her courtier is the greatest honor owe can possibly possess, for to serve the Queen of Heaven is already to reign there, and to live under her commands is more than to govern.   JOHN DAMASCENE

At the command of Mary all obey, even God.   BERNARDINO OF SIENA

ANGELS

Angels are spiritual beings, superior in nature to man, who were created by God to serve him obediently and to act as his messengers to us. Angels are more popular among us than are saints and appear in the plots of our motion pictures with regularity. Possibly their ecumenical character has something to do with their eternal appeal; all religions have angels or their analog, and the best-known angels, Gabriel, Raphael, and Michael, are called by name in the three monotheistic religions.

"He shall give his angels charge over thee, to keep thee in all thy ways. They shall bear thee up in their hands lest thou dash thy foot against a stone" (Psalms 91:11, 12). Several saints, including Gemma Galgani and Frances of Rome, are recorded to have seen and conversed with their guardian angels daily.

You should be aware that the word "angel" denotes a function rather than a nature. Those holy spirits of heaven have always been spirits, but they can only be called angels when they deliver some message. Those who deliver messages of lesser importance are called angels, while those who proclaim messages of supreme importance are called archangels.

GREGORY THE GREAT

Among the angels there is likewise this graded subordination of one hierarchy to another, and in the heavenly bodies and all their movements an orderly and close connection and interrelation is kept, all movement coming from the one Supreme Mover in perfect order, step by step, to the lowest.

IGNATIUS OF LOYOLA

There is a spiritual life that we share with the angels of Heaven and with the divine spirits, for like them we have been formed in the image and likeness of God.

LAWRENCE OF BRINDISI

Those who are in the first rank, and have as it were wings on their breasts and carry before them faces like the faces of men, in which the countenances of men appear as in pure water: these are angels, expanding the desires of a profound intellect like wings, not that they have wings like birds, but that they perform the will of God quickly in their desires, as a man flies quickly in his thoughts. HILDEGARD

So great was my joy in Him that I took no heed of looking at the angels and the saints, because all their goodness and all their beauty was from Him and in Him; He was the whole and Supreme Good, with all beauty, and so great a joy had I in His words that I paid no heed to any creature.
BL. ANGELA OF FOLIGNO

If we saw an angel clearly, we should die of pleasure. BRIDGET OF SWEDEN

If you find it impossible to pray, hide behind your good Angel and charge him to pray in your stead. JOHN VIANNEY

Since God often sends us inspirations by means of his angels, we should frequently return our aspirations to him by means of the same messengers.
FRANCIS DE SALES

Are the guardian angels standing by us, or are they still at a great distance? For until they come close to us, our efforts are vain and futile. Our prayer has neither the power of access nor the wings of purity to take it and bring it to the Lord. JOHN CLIMACUS

As Christ was pleased to be comforted by an angel, so was it necessary that the Virgin should be encouraged by one. PETER CHRYSOLOGUS

My guardian angel came and asked me what was the matter; I asked him to stay with me all night, and he said, "But I must sleep!" "No," I replied, "The angels of Jesus do not sleep!" "Nevertheless," he rejoined, smiling, "I ought to rest. Where shall you put me?" I begged him to remain near me. I went to bed; after that he seemed to spread his wings and come over my head. In the morning he was still there. GEMMA GALGANI

I tell you truly, my beloved, that our carelessness and our humiliation and our turning aside from the way are not a loss to us only, but they are a weariness for the angels and for all the saints in Christ Jesus. Our humiliation gives grief to them all, and our salvation gives joy and refreshment to them all. ANTHONY OF EGYPT

When tempted, invoke your angel. He is more eager to help you than you are to be helped! Ignore the devil and do not be afraid of him: he trembles and flees at your guardian angel's sight. JOHN BOSCO

THE WORLD

Because it is God's creature, the world can echo his glory, although the fall has left even nature suspect. Human society is, however, in every age, worse than ever. The world of creation is totally subject to God's will, but the world of sin, which is the product of man's self-will, is tainted. Jesus spoke of the saints and the world of sin when he said, "They are not of the world, even as I am not of the world" (John 17:16). Even if the world were much better than it is, or ever has been, the saint would still say, as Theophane Venard did to his executioners, "My heart is too big for this world, and nothing here can satisfy it."

I beg of you for the love and reverence of God our Lord to remember the past, and reflect not lightly but seriously that the earth is only earth.

IGNATIUS OF LOYOLA

When I say "the world," I mean the corrupt and disordered life led in the world, the damnable spirit that reigns over the world, the perverse sentiments and inclinations which men of the world follow, and the pernicious laws and maxims by which they govern their behavior. Christ looks upon the world as the object of His hatred and His curse, and as something he plans and desires to burn in the day of His wrath.   JEAN EUDES

Who love thee know thee not; who look down on thee, understand thee. Thus art thou not truthful, but deceitful—thou makest thyself true; and showest thyself false.   COLUMBANUS

Where now are the friends, the make-believes, the followers of the fashion? Where the suppers and feasts? Where the swarms of hangers-on? The strong wine decanting all day long, the cooks and the daintily dressed table, the attendants on greatness and all the words and ways they used to please?

They were all night and dreaming: now it is day and they are vanished. They were spring flowers, and, spring over, they all are faded together. They were a shadow, and it has traveled on beyond. They were smoke, and it has gone out in the air. They were bubbles and are broken. They were cobweb, and are swept away. And so this spiritual refrain is left again and again for us to sing: vanity of vanities, all is vanity.

JOHN CHRYSOSTOM

All the ways of this world are as fickle and unstable as a sudden storm at sea. BEDE THE VENERABLE

Worldly people often purchase Hell at a very dear price by sacrificing themselves to please the world. BL. HENRY SUSO

During eight-and twenty years of prayer, I spent more than eighteen in that strife and contention which arose out of my attempts to reconcile God and the world. TERESA OF AVILA

Not only pagan literature, but the whole sensible appearance of things is the lotus flower; so men forget their own land, which is God, the country of us all. PAULINUS OF NOLA

The truly wise man did well to call the falsely wise fools, for the wisdom of this world is foolishness with God. BERNARD OF CLAIRVAUX

We ought to love what Christ loved on earth, and to set no store by those things which he regarded as of no account. JOHN VIANNEY

Have no intercourse with the people in the world. Little by little you will get a taste for their habits, get so drawn into conversation with them that you will no longer be able out of politeness to refrain from applauding their discourse, however pernicious it may be, and it will lead you away into unfaithfulness. JEAN BAPTISTE DE LA SALLE

The new chief justice came to see me yesterday and asked all sorts of questions. He remarked that the happiness of the next world is doubtful, whereas present joys are certain. I told him that I could find nothing on earth that gives real happiness. Riches and pleasures do not satisfy very long; on the contrary, they bring along their own evils. I told him that my heart was too big for this world and nothing here could satisfy it.

BL. THEOPHANE VENARD

You have gone up into the mountain of sacrifice, while I still dwell in the valley of care, and have spent almost all my life for others. You take the wings of contemplation and soar above all this, but I am so stuck in the glue of concern for other people that I cannot fly. Who will give me wings like a dove so that I can fly away and find rest?   PETER MARTYR

If I could live in a tiny dwelling on a rock in the ocean, surrounded by swelling waves, cut off from the knowledge and the sight of all, I would still not be free from the cares of this fleeting world nor from the fear that somehow the love of money would come and snatch me away.

CUTHBERT OF LINDISFARNE

It is a great sorrow for a soul that wishes to live far from the pomps and vanities to return to the world, to put up with idle and insipid conversations instead of talking to God alone, to open one's eyes to see nothing but the earth instead of visions of heaven. A hard sacrifice indeed.

GEMMA GALGANI

It is little use expecting anything from the mighty ones of this world—for the most part they leave the poor to their poverty, and mean and ungenerous as they are, turn a deaf ear to the cry of those who are weak and helpless.   BL. PLACID RICCARDI

Alas, Monseigneur, I am obliged to live amongst secular people and cannot avoid being mixed up in the affairs of the world. I am polluted by the dust which springs up round the people of the age.

LYDWINA OF SCHIEDAM, *to an English Bishop*

Let the world indulge its madness, for it cannot endure and passes like a shadow. It is growing old, and I think, is in its last decrepit stage. But we, buried deep in the wounds of Christ, why should we be dismayed?

PETER CANISIUS

Our labor here is brief, but the reward is eternal. Do not be disturbed by the clamor of the world, which passes like a shadow. Do not let the false delights of a deceptive world deceive you.   CLARE OF ASSISI

Young men, you and I set little store by earthly existence. Things that merely improve this life have no true value for us; they are not what we call "the real thing." Good family, athletic valour, a handsome face, tall stature, men's esteem, dominion over others—none of these are important in our eyes or a petition fit for prayer; it is not our way to pay court to those who can boast them. Our ideals soar far above all that.

BASIL THE GREAT

I think of the joy, the love, the peace of those who dedicate themselves to God alone. Then I call to mind the misfortunes, the tortures, the remorse, and the disturbance of those who long for the things of this world with so much solicitude and ardor. Then, with all my strength I call on all men who dwell on this earth to lift themselves with me to God, in order to bless and praise him. BL. HENRY SUSO

Just as the cloud of unknowing lies above you, between you and your God, so you must fashion a cloud of forgetting beneath you, between you and every created thing. ANONYMOUS, *The Cloud of Unknowing*

Let us not esteem worldly prosperity or adversity as things real or of any moment, but let us live elsewhere, and raise all our attention to Heaven; esteeming sin as the only true evil, and nothing truly good but virtue which units us to God. GREGORY NAZIANZEN

Perfection does not consist in not seeing the world, but in not having a taste or relish for it. FRANCIS DE SALES

Those things that are more scorned and shunned by worldly men are honored and valued by God and His saints. And those that are more loved and embraced and honored by worldly men are more hated and shunned and scorned by God and his saints. Men hate everything that should be loved and love what should be hated. BL. GILES OF ASSISI

As long as we have peace with the natures of this world, we are enemies of God and of His angels, and of all His saints. ANTHONY OF EGYPT

We must look upon all things of this world as none of ours, and not desire them. This world and that to come are two enemies. We cannot therefore be friends to both; but we must resolve which we would forsake and which we would enjoy. CLEMENT OF ROME

Our thoughts ought instinctively to fly upwards from animals, men, and natural objects to their Creator. If things created are so full of loveliness, how resplendent with beauty must be He who made them! The wisdom of the Worker is apparent in His handiwork. ANTHONY OF PADUA

A strong, resolute soul can live in the world without being infected by any of its moods, find sweet springs of piety amid its salty waves, and fly through flames of earthly lusts without burning the wings of its holy desires for a devout life. FRANCIS DE SALES

Every creature in the world will raise our hearts to God if we look upon it with a good eye. FELIX OF CANTALICE

Living, as I do, in the whirlwind of the court, I am hardly ever able to reach, or, to be quite honest, even to see from afar, the tranquility of contemplation. I am so busy with Leah's morning shortsightedness and fruit-fulness that in my present position I cannot reach the beauty of Rachel to which I have aspired.   RAYMOND OF PEÑAFORT

The laws and maxima of Jesus are very mild and holy and reasonable. The standards of the world are laws and maxims of hell, and are diabolical, tyrannical, and finally unbearable.   JEAN EUDES

TIME AND ETERNITY

Eternity is God's time. He has no beginning, he never changes, and he will never end. God possesses the divine being in a constant, undivided now. Historical time is his creation. It had a beginning, at the creation, and will have an end, at the time of the last judgement. Angels and human beings will then dwell in God's time.

It is the prospect of eternity that makes temporal suffering and sacrifice completely logical and reasonable. Meanwhile, the church must exist within historical time. Our life in the world is seen by the saints as a brief exile from God's time. While the mystics dwell on the shores of eternity, the saints of action live in time and try to transform it, through fidelity to and in the present.

Were you to quote to me the proverbial old age of Tithonus or even of Methusulah, the oldest man in all history who attained, so they say, a thousand years save thirty, were you to pile up the years of mortal man and add them all together, why, I should laugh as at the babbling of babes when I consider Eternity—Eternity, everlasting and ageless, that human thought can never grasp nor encircle, any more than it can assign any end to an immortal soul. BASIL THE GREAT

The eyes of the world see no further than this life, as mine see no further than this wall when the church door is shut. The eyes of the Christian see deep into Eternity. JOHN VIANNEY

The first ideas I can remember date back to when I was five years old. When I went to bed, instead of sleeping—I have never been much of a sleeper—I used to think about eternity. I would think "forever, forever, forever." I would try to imagine enormous distances and pile still more distances on these and realize that they would never come to an end.
ANTHONY MARY CLARET

My brother and I were quite frightened to know, from our reading, that suffering and glory were forever. It befell us to speak about this many times and we liked to say, "forever, forever, forever!" Through my frequently saying this, Our Lord left me impressed, even in my childhood, with the pathway to truth.   TERESA OF AVILA

Sometimes I felt very lonely, depressed and ill. I'd often repeat to myself a line of poetry which brought peace and strength back into my soul; it runs, "Time's but a ship that bears thee, not thy home." That image appeals to me and helps me to bear this life of exile.   THÉRÈSE DE LISIEUX

I travel, work, suffer my weak health, meet with a thousand difficulties, but all these are nothing, for this world is so small. To me, space is an imperceptible object, as I am accustomed to dwell in eternity.

FRANCES XAVIER CABRINI

The entire life span of men is very brief when measured against the ages to come, so that all our time is nothing in comparison with eternal life. Everything in the world is sold for what it is worth, and someone trades an item for its equivalent. But the promise of eternal life is purchased for very little.   ATHANASIUS

Don't imagine that if you had a great deal of time you would spend more of it in prayer. Get rid of that idea! Again and again God gives more in a moment than in a long period of time, for his actions are not measured by time at all.   TERESA OF AVILA

Time and eternity are one and the same thing in God; and the temporal being of the creature in the nature and essence of God has no longer any diversity.   BL. HENRY SUSO

He observed that in saying *today* he was not counting the time passed, but as one always establishing a beginning, he endeavored each day to present himself as the sort of person ready to appear before God.

ATHANASIUS, *The Life of Anthony*

We must adapt the Society to the times, not the times to the Society.

IGNATIUS OF LOYOLA

It shows weakness of mind to hold too much to the beaten track through fear of innovations. Times change and to keep up with them, we must modify our methods.   MADELEINE SOPHIE BARAT

The past must be abandoned to God's mercy, the present to our fidelity, the future to divine providence.   FRANCIS DE SALES

HUMAN NATURE

For all the trouble that it gives, evidently human nature is the way God meant it to be. Angels were created as pure spirit, while man, a slightly inferior being, is both spirit and matter, made in the image of God, but also of dust. The odd and difficult combination of soul and body—the 'mind-body problem' with which philosophers continue to grapple—is intended by God to be our special dilemma.

Men go abroad to wonder at the height of mountains, at the huge waves of the sea, at the long courses of the rivers, at the vast compass of the ocean, at the circular motion of the stars, and they pass by themselves without wondering.   AUGUSTINE OF HIPPO

Our condition is most noble, being so beloved of the most high God that He was willing to die for our sake, which He would not have done if man had not been a most noble creature and of great worth.

BL. ANGELA OF FOLIGNO

It is not our body which feels, not our mind which thinks, but we, as single human beings, who both feel and think.   THOMAS AQUINAS

He who knows himself knows all men. Therefore it is written, 'He called all things out of nothingness into being.' Such statements refer to our intellectual nature, which is hidden in this body of corruption, but which did not belong to it from the beginning, and is to be freed from it. But he who can love himself, loves all men.   ANTHONY OF EGYPT

We who are slaves to Christ make our bodies serve and our minds govern, so that the flesh receives its orders and accompanies our will which is guided by Christ our Maker. The body derives steadfastness from the mind's courage, and the servant obeys in accordance with the disposition of the master.

PAULINUS OF NOLA

Raise up your heart after a fall, sweetly and gently, humbling yourself before God in the knowledge of your misery, and do not be astonished at your weakness, since it is not surprising that weakness should be weak, infirmity infirm, and frailty frail.   FRANCIS DE SALES

It is not possible for all things to be well, unless all men were good, which I think will not be this good many years.   THOMAS MORE

God is like a mother who carries her child in her arms by the edge of a precipice. While she is seeking all the time to keep him from danger, he is doing his best to get into it.   JOHN VIANNEY

Now true love and lasting friendship require certain dispositions; those of our Lord, we know, are absolutely perfect; ours, vicious, sensual, and thankless; and you cannot, therefore, bring yourselves to love Him as He loves you because you have not the disposition to do so.   TERESA OF AVILA

Bereft of their bodies, the souls of the blessed ones have neither the wish nor the power to reach their ultimate end. Therefore, until such time as their bodies are restored to them, souls cannot be absorbed into God with that fullness which is their loftiest, their perfect state. Neither would the spirit yearn once more for the fellowship of the flesh were it possible to reach the perfect condition of human nature in aught other way.

BERNARD OF CLAIRVAUX

I saw a body lying on the earth: which body showed heavy and fearful, and without shape and form, as it were a swilge stinking myre. And suddenly out of this body sprung a full fair creature, a little child full shapen and formed, swift and lively, and whiter than the lily, which sharply glided up into heaven. The swilge of the body betokeneth great wretchedness of our deadly flesh: and the littleness of the child betokeneth the cleanness and the pureness of our soul.   JULIAN OF NORWICH

Now to bring man to being, to make a living and rational creature, of bones, muscle, veins, and all the rest of man's economy, which as yet did not exist, was a task far harder and more incredible than to restore this creature to life after it had been re-dissolved into the earth.   IRENAEUS

By what rule or manner can I bind this body of mine? By what precedent can I judge him? Before I can bind him he is set loose, before I can condemn him I am reconciled to him, before I can punish him I bow down to him and feel sorry for him. How can I hate him when my nature persuades me to love him? How can I break away from him when I am bound to him forever? How can I escape from him when he is going to rise with me?   JOHN CLIMACUS

Whether in this life, or in death, or in the final resurrection, the body availeth much to the soul that loveth the Lord. In the first case, it produces the fruit of penitence; in the second, the boon of rest; and in the third, the last condition of beatitude. The soul is right in deeming that since the body is of service to her in every state, it too should have a part in perfection.

BERNARD OF CLAIRVAUX

GRACE

Grace is an unmerited gift bestowed on the human race by the benevolence of God and includes all the supernatural charisms of the Holy Spirit. It is a paradoxical concept which can create contradiction, confusion, and a sense of great unfairness. The Calvinist doctrine of election and predestination is a famous example of the difficulty inherent in the idea of grace which emphasizes that some depraved creatures are chosen for salvation, while others, equally unworthy, are left to perish. Christ repeatedly told us that fairness is not the issue, as in the parables of the lost sheep, the prodigal son, and the laborers in the vineyard.

The saints take their stand in the humble acknowledgment that they are quite unworthy of grace, that they hunger desperately for it, and that they are terribly grateful for it, more than willing to suffer to get more of it, and still, as Augustine says, less willing to accept grace than God is to give it.

We must seldom speak of predestination; and if it should happen that we do so, it must not be in such a way that people can say, "If my eternal fate is fixed, whether I do good or well, it will only be what God has decided"; which too frequently leads to the neglect of good works and of all the means of salvation.   IGNATIUS OF LOYOLA

Here there begins an eternal hunger, which shall never more be satisfied; it is an inward craving and yearning of the created spirit for the loving power of an uncreated God. And since the spirit longs for fruition and is invited and urged thereto by God, it must continually desire its fulfillment. Behold, now begins an eternal craving and insatiable longing.

BL. JAN VAN RUYSBROECK

Man, blinded and bowed, sits in darkness and cannot see the light of heaven unless grace with justice comes to his aid.   BONAVENTURE

Since, by assenting to what belongs to faith, man is raised above his nature, this must needs come to him from some supernatural principle moving him inwardly, and this is God. Therefore faith, as regards the assent which is the chief act of faith, is from God moving inwardly by grace.

THOMAS AQUINAS

Reawaken my soul by the grace of Your love, since it is Your commandment that we love You with all our heart and strength—and no one can fulfill that commandment without Your help.   BL. JOHN OF ALVERNA

Our Lord God showed that it is full great pleasance to Him that a silly soul come to Him nakedly and plainly and homely. For this is the kind yearnings of the soul, by the teaching of the Holy Ghost.

BL. JULIAN OF NORWICH

If God treated you according to your merits He would drive you out of His sight and destroy you when you presented yourself before Him. Therefore, when He gives you grace, you must not think that He gives it to you in answer to your prayers, but rather than He gives it to His son Jesus Christ by virtue of His prayers and merits.   JEAN EUDES

Nothing is sweeter than the calm of conscience, nothing safer than purity of soul, which yet no one can bestow on himself because it is properly the gift of another.   COLUMBANUS

He who aspires to the grace of God must be pure, with a heart as innocent as a child's. Purity of heart is to God like a perfume sweet and agreeable.

NICHOLAS OF FLUE

"My delight is to be with the children of men." I must not infringe the rights of your freedom; but I will transform you in Myself, since you wish it, and make you one with Me by making you share in My perfection, particularly in My tranquillity and My peace.   CATHERINE OF SIENA

How should Our Lord fail to grant his graces to him who asks for them from his heart when He confers so many blessings even on those who do not call on Him? Ah, He would not so urge and almost force us to pray to Him if He had not a most eager desire to bestow His graces on us.

JOHN CHRYSOSTOM

God is more anxious to bestow His blessings on us than we are to receive them.   AUGUSTINE OF HIPPO

Perfect correspondence to his grace consists in a strong, deep, interior sorrow.
ANONYMOUS, *The Cloud of Unknowing*

Mine is indeed a fearful responsibility. The mere thought of it makes me tremble. What a terrible account I shall have to render to God for all the graces that he has bestowed on us for the progress of our Congregation.

JOHN BOSCO

There is no one who during this mortal life can properly judge how far he is an obstacle and to what extent he resists the workings of God's grace in his soul. IGNATIUS OF LOYOLA

My God, how good you are! How rich in mercy you have been to me! If you had given others the graces you have given me, they would have co-operated with them so much more. Mercy, Lord: I'll begin to be good from now, with the help of your grace. ANTHONY MARY CLARET

Our Lord takes pleasure in doing the will of those who love him.

JOHN VIANNEY

Those who imagine they can attain to holiness by any wisdom or strength of their own will find themselves after many labours, and struggles, and weary efforts, only the farther from possessing it, and this in proportion to their certainty that they of themselves have gained it. BL. JOHN OF AVILA

Why did Christ refer to the grace of the Spirit under the name of water? Because through water all vegetables and animals live. Because the water of rain comes down from heaven, and though rain comes down in one form its effects take many forms. Yea, one spring watered all of paradise, and the same rain falls on the whole world, yet it becomes white in the lily, red in the rose, purple in the violet. CYRIL OF JERUSALEM

Our Lord and Savior lifted up his voice and said with incomparable majesty: "Let all men know that grace comes after tribulation. Let them know that without the burden of afflictions it is impossible to reach the height of grace. Let them know that the gifts of grace increase as the struggles increase." ROSE OF LIMA

The Holy Spirit rests in the soul of the just like the dove in her nest. He hatches good desires in a pure soul, as the dove hatches her young.

JEAN VIANNEY

O Holy Ghost, that givest grace where Thou wilt, come into me and ravish me to Thyself. The nature that Thou didst make, change with honeysweet gifts, that my soul, filled with Thy delightful joy, may despise and cast away all the things of this world, that it may receive ghostly gifts, given by Thee, and going with joyful songs into infinite light may be all melted in holy love. BL. RICHARD ROLLE

PRAYER

The most consistent theme of saints' utterances on prayer is simplicity and striving toward loving union with God in spite of all the complications and distractions that inevitably arise. This is as true of the active "worldly" saints as of those whose primary vocation is prayer and contemplation.

The saints give a wealth of sometimes conflicting advice on how to pray, reflecting different needs, temperaments, and vocations. The contemplatives, like John of the Cross and the anonymous author of the *Cloud of Unknowing*, are more concerned with the nuances of the soul's movement through different stages on the journey toward union with God than are God's soldiers in the world, like Ignatius of Loyola or Anthony Mary Claret, who just give us the best advice they can on how to stay close to God in the midst of tumult and distraction, based on their own amazing experience.

For me, prayer means launching out of the heart towards God; it means lifting up one's eyes, quite simply, to Heaven, a cry of grateful love from the crest of joy or the trough of despair; it's a vast, supernatural force which opens out my heart, and binds me close to Jesus.   THÉRÈSE DE LISIEUX

Of course it is laudable to reflect upon God's kindness and to love and praise him for it; yet it is far better to let your mind rest in the awareness of him in his naked existence and to love and praise him for what he is in himself.   ANONYMOUS, *The Cloud of Unknowing*

I don't say anything to God. I just sit and look at him and let him look at me.   OLD PEASANT OF ARS

All that the beginner in prayer has to do—and you must not forget this, for it is very important—is to labor and be resolute and prepare himself

with all possible diligence to bring his will into conformity with the will of God.
                                                                                    TERESA OF AVILA

After I enter the chapel I place myself in the presence of God and I say to Him. "Lord, here I am; give me whatever you wish." If he gives me something, then I am happy and I thank him. If he does not give me anything, then I thank him nonetheless, knowing, as I do, that I deserve nothing. Then I begin to tell him of all that concerns me, my joys, my thoughts, my distress, and finally, I listen to him.   CATHERINE LABOURÉ

You may laugh, Jesus, but it's no laughing matter for me. Listen, Jesus: will you do this for me? Come on, say yes; if not some harm will come of it. Don't think of what I deserve, but of the merits of those who pray for me.
                                                                                    GEMMA GALGANI

Countless numbers are deceived in multiplying prayers. I would rather say five words devoutly with my heart than five thousand which my soul does not relish with affection and understanding.   EDMUND THE MARTYR

Vocal prayer suits me better than strictly mental prayer, thank God. At each word of the Our Father, Hail Mary, and Glory, I glimpse an abyss of goodness and mercy. Our Lord has granted me the grace of being very attentive and fervent when I say these; in His goodness he also grants me many graces during mental prayer, but in vocal prayer I have a deeper awareness.   ANTHONY MARY CLARET

Inquire and meditate on all the ways of the Passion and the Cross. And even if thou art not able to do this with thine heart, at least with thy mouth shalt thou earnestly and diligently repeat those things which belong unto the said Passion, because when a thing is ofttimes spoken with the mouth it doth in the end impart warmth and fervor into the heart.

                                                            BL. ANGELA OF FOLIGNO

Learn to abide with attention in loving waiting upon God in the state of quiet; give no heed to your imagination, nor to its operations, for now, as I have said, the powers of the soul are at rest and are not exercised except in the sweet and pure waiting of love.   JOHN OF THE CROSS

For perfected souls every place is to them an oratory, every moment a time for prayer. Their conversation has ascended from earth to heaven—that is to say they have cut themselves off from every form of earthly affection and sensual self-love and have risen alone into the very height of Heaven.

                                                            CATHERINE OF SIENA

No matter how much our interior progress is ordered, nothing will come of it unless by divine aid. Divine aid is available to those who seek it from their hearts, humbly and devoutly; and this means to sigh for it, in this valley of tears, through fervent prayer.   BONAVENTURE

We are told how the monks of Egypt prayed very frequently but very briefly. Their prayer was sudden and ejaculatory so that the intense application so necessary in prayer should not vanish or lose its keenness by a slow performance.   AUGUSTINE OF HIPPO

It is an old custom of the servants of God to have some little prayer ready and to be frequently darting them up to Heaven during the day, lifting their minds to God out of the mire of this world. He who adopts this plan will get great fruits with little pains.   PHILIP NERI

Aspire to God with short but frequent outpourings of the heart; admire his bounty; invoke his aid; cast yourself in spirit at the foot of His cross; adore his goodness; treat with Him of your salvation; give Him your whole soul a thousand times in the day.   FRANCIS DE SALES

All our life is like a day of celebration for us; we are convinced, in fact, that God is always everywhere. We work while singing, we sail while reciting hymns, we accomplish all other occupations of life while praying.
                                   CLEMENT OF ALEXANDRIA

The prayer of the mind is not perfect until he no longer realizes himself or the fact that he is praying.   ANTHONY OF EGYPT

Souls without prayer are like people whose bodies or limbs are paralyzed: they possess feet and hands but they cannot control them. In the same way there are souls so infirm and so accustomed to busying themselves with outside affairs that they seem incapable of entering into themselves at all.
                                         TERESA OF AVILA

It is indeed essential for a man to take up the struggle against his thoughts if the veils woven from his thoughts and covering up his intellect are to be removed, thus enabling him to turn his gaze without difficulty towards God and to avoid following the will of his wandering thoughts.
                                       AMMONAS THE HERMIT

Painful and difficult prayer is more pleasing to God than one which is easy and tranquil. The grief and pain of one who tries to pray in vain, lamenting his inability to do so, makes him a victor in God's sight and obtains for him abundant graces.   BL. HENRY SUSO

Usually prayer is a question of groaning rather than speaking, tears rather than words. For He sets our tears in His sight, and our groaning is not hidden from Him who made all things by His Word and does not ask for words of man.   AUGUSTINE OF HIPPO

When we are linked by the power of prayer, we, as it were hold each other's hand as we walk side by side along a slippery path; and thus by the bounteous disposition of charity, it comes about that the harder each one leans on the other, the more firmly we are riveted together in brotherly love.					GREGORY THE GREAT

I know that soldiers have a lot to endure, and to endure in silence. If upon rising they would only take the trouble to say to our Lord every morning this tiny phrase: "My God, I desire to do and to endure everything today for love of Thee," what glory they would heap up for eternity! Why, a soldier who did that and was as loyal as possible to his Christian duties would earn as much reward as any cloistered monk.

BERNADETTE SOUBIROUS, *to her brother*

Do not fail to apply yourself to whatever inspires the most devotion in you. The most beneficial prayer will be the one which moves your heart in the most beneficial way.					BL. JORDAN OF SAXONY

What our Creator and Redeemer puts into the heart and what moves it most to piety is the best.					BL. JOHN OF AVILA

Prayer is a fragrant dew, but we must pray with a pure heart to feel this dew. There flows from prayer a delicious sweetness, like the juice of very ripe grapes. Troubles melt away before a fervent prayer like snow before the sun. To approach God one should go straight to him, like a ball from a cannon. Prayer disengages our soul from matter; it raises it on high, like the fire that inflates a balloon. The more we pray, the more we wish to pray. Like a fish which at first swims on the surface of the water, and afterwards plunges down and is always going deeper, the soul plunges, dives, and loses itself in the sweetness of conversing with God. Prayer is the holy water that by its flow makes the plants of our good desires grow green and flourish, that cleanses our souls of their imperfections, and that quenches the thirst of passion in our hearts.					JOHN VIANNEY

However great may be the temptation, if we know how to use the weapon of prayer well we shall come off conquerors at last, for prayer is more powerful than all the devils. He who is attacked by the spirits of darkness needs only to apply himself vigorously to prayer and he will beat them back with great success.					BERNARD OF CLAIRVAUX

Lift up your heart to him, at the beginning of every action, somewhat like this: "O Jesus, with all my power I renounce myself, my own mind, my own will, and my self-love, and I give myself all to Thee and to Thy holy spirit and Thy divine love. Draw me out of myself and direct me in this action according to Thy holy will."					JEAN EUDES

Our chiefest labor in prayer must be to inflame and set our hearts on fire, with this fervency of charity, and then as it were to spin out our prayer, so long until we have attained unto this end. But when through weariness of our frail body we find this heat and fervor in us to grow cold, then must we desist and pray no longer, but presently apply ourselves to some other works of virtue. JOHN FISHER

Up to this time nobody had taught me the act of mental prayer; I should have liked to know about it but my sister was satisfied with my spiritual progress as it was and kept me to vocal prayer instead. One day, one of my teachers at the Abbey asked me what I did with myself on holidays, when I was left to my own devices. I told her that I got behind my bed, where there was an empty space in which you could shut yourself away with the curtains, and there . . . well I used to think. "Think about what?" she asked. "Oh," I said, "About God and about life and eternity; you know, I just think." THÉRÈSE DE LISIEUX

Build yourself a cell in your heart and retire there to pray.
CATHERINE OF SIENA

Great talent is a gift of God, but it is a gift which is by no means necessary in order to pray well. This gift is required in order to converse well with men; but it is not necessary in order to speak well with God. For that, one needs good desires, and nothing more. JOHN OF THE CROSS

What am I lying here for? God will certainly have heard the prayers of so many good men. Fetch me my shoes and stick! CUTHBERT OF LINDISFARNE

CONVERSION

_____

The saints discuss many types of conversion. In a sense this is an artificial category, since many stages on the ascent to union with the divine can be called conversions from the lower to the higher. Conversion is a never-ending process, but there is also a very dramatic 'road to Damascus' experience of turning away from one's past and toward God, which is what we usually refer to when using the word.

_____

The Lord pursued me for a long time. He put me, as it were, into prison in order to force me to contemplate him and speak to him. He deprived me of everything that I might go and prostrate myself at his feet; but invariably I again attached myself to nothingness in order to shun the abyss of love that Jesus had in store for me.   PETER JULIAN EYMARD

He would never come and knock at the door unless he wished to enter; if he does not always enter, it is we who are to blame.   AMBROSE

If the heart of man is not lifted up, this is from no defect on the part of him who draws it, who as far as he is concerned never fails, but from an impediment caused by him who is being drawn.   THOMAS AQUINAS

There are souls which at first were hard of heart and persisted in the works of sin; and somehow the good God in his mercy sends upon such souls the chastisement of affliction, till they grow weary, and come to their senses, and are converted, and draw near, and enter into knowledge, and repent with all their heart, and they also attain the true manner of life.

ANTHONY OF EGYPT

How sweet did it at once become to me to be without the sweetness of those baubles! What I feared to be parted from, it was now a joy to part with for thou didst cast them from me, thou the true and richest sweetness.

Thou didst cast them forth, and in their place didst substitute thyself, sweeter than all delight. AUGUSTINE OF HIPPO

Have we not had the brightest examples of the conversions of the human passions for our Christ in the direction of love? Did a Magdalen, a Paul, a Constantine, an Augustine become mountains of ice after their conversion? Quite the contrary. We should never have had these prodigies of conversion and marvelous holiness if they had not changed the flames of human passion into volcanoes of immense love of God. FRANCES XAVIER CABRINI

When I had drunk the spirit from Heaven, and the second birth had restored me so as to make me a new man then immediately in a marvelous manner doubts began to be resolved, closed doors to be opened, dark places to be light; what before was difficult now seemed easy. CYPRIAN

The moment I realized that God existed, I knew I could not do otherwise than to live for Him alone. VEN. CHARLES DE FOUCAULD

You are standing in front of God and in the presence of the hosts of angels. The Holy Spirit is about to impress his seal on each of your souls. You are about to be pressed into the service of the great king.
CYRIL OF JERUSALEM

One night, as he lay awake, he saw clearly the likeness of our Lady with the holy Child Jesus, at the sight of which he received most abundant consolation for a considerable interval of time. He felt so great a disgust with his past life, especially with its offenses of the flesh, that he thought all such images which had formerly occupied his mind were wiped out. And from that hour until August of 1553, when this is being written, he never again consented to the least suggestion of the flesh. IGNATIUS OF LOYOLA

When I had come by ill luck to Ireland—well, every day I used to look after sheep and I used to pray often during the day, the love of God and the fear of him increased more and more in me and my faith began to grow and my spirit to be stirred up, so that in one day I would say as many as a hundred prayers and I used to rise at dawn for prayer, in snow and frost and rain because the Spirit was glowing in me. PATRICK OF IRELAND

Heretics are to be converted by an example of humility and other virtues far more readily than by any external display or verbal battles. So let us arm ourselves with devout prayers and set off showing signs of genuine humility and go barefooted to combat Goliath. DOMINIC

The priest seen only in the pulpit is doing his duty, but let him speak a word at recreation and he becomes a friend. How many changes of heart spring from a few chance words spoken in a boy's ear at play!   JOHN BOSCO

One Indian was strikingly converted. As he lay on his death bed he spoke of a previous illness during which he had thought he was dying and he said aloud: "I then saw the Author of Life and He said to me, 'Go back, your hour is not yet!' But I know that this time I *shall* go to the Author of Life." Francis, a Christian Iroquois present, said to him, "The Author of Life probably sent you back so that you might have water poured on your head." The dying Sioux made answer: "Indeed, I think it was precisely for that I was told to return to life."   BL. PHILIPPINE DUCHESNE

A few acts of confidence and love are worth more than a thousand "who knows? who knows?" Heaven is filled with converted sinners of all kinds and there is room for more.   JOSEPH CAFASSO

When grace draws a man to contemplation it seems to transfigure him even physically so that though he may be ill-favored by nature, he now appears changed and lovely to behold. His whole personality becomes so attractive that good people are delighted to be in his company, strengthened by the sense of God he radiates.   ANONYMOUS, *The Cloud of Unknowing*

I must also pray, she said, for the conversion of sinners. I asked her many times what she meant by that, but she only smiled.

BERNADETTE SOUBIROUS

Finally I beheld God in spirit during Mass, at about the time of the elevation of the Body of Christ. After this vision there remained unto me an indescribable sweetness and great joy which I do think will never fail me all the days of my life.   BL. ANGELA OF FOLIGNO

PERFECTION

Moral perfection consists in becoming like Christ. For us, full perfection can be attained only after resurrection and attainment of the vision of God. Relative perfection on this earth is to be measured by the practice of the theological virtues of faith, hope, and charity. The Christian is obliged to strive for perfection here and now through the practice of obedience to God's law and will.

François Mauriac has said, "The road to perfection skirts the abyss of despair. To the very end, despair remains the temptation of those who have not retreated in the face of Christ's command: 'Be you therefore perfect as also your heavenly father is perfect.'" Not to despair, to persevere in the attempt, is the definition of that heroic virtue which marks the saint.

God has not called his servants to a mediocre, ordinary life, but rather to the perfection of a sublime holiness since he said to his disciples: "Be ye perfect as your heavenly Father is perfect."   BL. HENRY SUSO

We have dared to attempt our work of perfection, relying not on our own deeds and strength but on the power and mercy of God. Since He is almighty, He can complete in us the work of His perfection. When He has deigned to lay the foundation and to begin the first scaffolding, He can construct it according to His measurements and complete it by roofing it.   PAULINUS OF NOLA

Just as a mother is able to offer food to an infant, but the infant is not yet able to receive food unsuited to its age, so God could have offered perfection to man at the beginning, but man, being yet an infant, could not have taken it.   IRENAEUS

Our perfection doth certainly consist in knowing God and ourselves.
          BL. ANGELA OF FOLIGNO

Christian perfection consists in three things: praying heroically, working heroically, and suffering heroically.   ANTHONY MARY CLARET

Man's salvation and perfection consist in doing the will of God, which he must have in view in all things and at every moment of his life: The more he accomplishes this divine will, the more perfect he will be.

PETER CLAVER

If you do that which I am going to tell you now, you will have reached a consummate perfection and nothing will be wanting in you. It is the attainment of an ardently desired and perseveringly sought disposition of the soul in which you are so closely united with Me and your will so conformed to My perfect will that you will never wish not only evil, but even the good that I do not wish.   CATHERINE OF SIENA

Therefore with mind entire, faith firm, courage undaunted, love thorough, let us be ready for whatever God wills; faithfully keeping His commandment, having innocence in simplicity, peaceableness in love, modesty in lowliness, diligence in ministering, mercifulness in helping the poor, firmness in standing for truth, and sterness in keeping of discipline.

BEDE THE VENERABLE

There are certain souls who desire to arrive at perfection all at once, and this desire keeps them in constant disquiet. It is necessary first to cling to the feet of Jesus, then to kiss his sacred hands, and at last you may find your way into his divine heart.   ALPHONSUS LIGUORI

The work of purging the soul neither can nor should end except with our life itself. We must not be disturbed at our imperfections, since for us perfection consists in fighting against them. How can we fight against them unless we see them, or overcome them unless we face them?

FRANCIS DE SALES

A glad spirit attains to perfection more quickly than any other.

PHILIP NERI

In all your deeds and words you should look upon this Jesus as your model. Do so whether you are walking or keeping silence, or speaking, whether you are alone or with others. He is perfect, and thus you will be not only irreprehensible, but praiseworthy.   BONAVENTURE

He belongs to you, but more than that, he longs to be in you, living and ruling in you, as the head lives and rules in the body. He wants his breath to be in your breath, his heart in your heart, and his soul in your soul, so that you may indeed "Glorify God and bear him in your body, that the life of Jesus may be made manifest in you."   JEAN EUDES

Thou shalt die to thyself so utterly as not to go to sleep at night until thou hast sought out thy tormentor and, as far as possible, calmed his angry heart with thy sweet words and ways; for with such meek lowliness thou wilt take from him his sword and knife, and make him powerless in his ill will. See, this is the old perfect way which the dear Christ taught his disciples when he said, "Behold I send you as lambs among wolves."

BL. HENRY SUSO

Let your intentions in the fulfillment of your duties be so pure that you reject from your actions every other object but the glory of God and the salvation of souls.   ANGELA MERICI

FAITH AND HOPE

The theological virtues of faith, hope, and love (charity) are supernatural gifts that lead the soul to God.

Faith consists in adherence, through grace, to a truth revealed by God because of the authority of God rather than because of the evidence given. "Blessed are they who have not seen and yet have believed" (John 20:29).

The virtue of hope has the possession of God and eternal happiness as its object. Its grounds are God's goodness, his power, his faithfulness to his promises, and, most specifically, the resurrection of Christ. Without hope, faith is weakened or disappears.

Faith is the proof of what cannot be seen. What is seen gives knowledge, not faith. GREGORY THE GREAT

We should submit our reason to the truths of faith with the humility and simplicity of a child. ALPHONSUS LIGUORI

It is because of faith that we exchange the present for the future.
FIDELIS OF SIGMARINGEN

We are not justified through ourselves or through our own wisdom or understanding or piety, or through actions done in holiness of heart, but through faith, for it is through faith that Almighty God has justified all men that have been from the beginning of time. CLEMENT OF ROME

Not only in actions but in faith also has God preserved man's free and unconstrained choice. He says, "Let it happen to you according to your faith," thus showing that faith is something which a man has as his own, just as he has his own power of decision. IRENAEUS

Faith comes from the disposition of the soul, but dialectic comes from the skill of those who construct it. Therefore for those in whom the action of faith is present, the demonstration through arguments is unnecessary, or even useless. What we perceive through faith you attempt to establish by arguments, and often you cannot even express that which we see, so it is clear that insight through faith is better and more secure than your sophistic conclusions. ANTHONY OF EGYPT

The mysteries and secrets of the kingdom of God first seek for believing men, that they may make them understanding. For faith is understanding's step, and understanding faith's attainment. AUGUSTINE OF HIPPO

The divine clemency has made this salutary commandment, that even some things which reason is able to investigate must be held by faith; so that all may share in the knowledge of God easily, and without doubt or error. THOMAS AQUINAS

O man, believe in God with all your might, for hope rests on faith, love on hope, and victory on love; the reward will follow victory, the crown of life the reward, but the crown is the essence of things eternal. NICHOLAS OF FLÜE

I long to understand in some degree thy truth, which my heart believes and loves. For I do not seek to understand that I may believe, but I believe in order to understand. For this also I believe, that unless I believed, I should not understand. ANSELM OF CANTERBURY

We must take care lest, by exalting the merit of faith, without adding any distinction or explanation, we furnish people with a pretext for relaxing in the practice of good works. IGNATIUS OF LOYOLA

If we never look at Him or think of what we owe Him and of the death which He suffered for our sakes, I do not see how we can get to know Him or do good works in His service. For what can be the value of faith without works, or works which are not united with the merit of our Lord Jesus Christ? And what but such thoughts can arouse us to love this Lord? TERESA OF AVILA

O how glorious our Faith is! Instead of restricting hearts, as the world fancies, it uplifts them and enlarges their capacity to love, to love with an almost infinite love, since it will continue unbroken beyond our mortal life. THÉRÈSE DE LISIEUX

Faith is a divine and celestial light, a participation in the eternal, inaccessible light, a beam radiating from the face of God. Although it is true that faith is accompanied by obscurity and permits you to behold God, not clearly as He is seen in Heaven, but as through a cloud, darkly, nevertheless faith does not debase His supreme greatness to fit the capacity of your mind, as does science.   JEAN EUDES

Faith is the unshaken stance of the soul and is unmoved by any adversity. The believing man is not one who thinks God can do all things, but one who trusts that he will obtain everything. Faith is the agent of things un-hoped for, as the thief proved. The mother of faith is hard work and an upright heart; the one builds up belief, the other makes it endure.

JOHN CLIMACUS

Gladly endure whatever goes against you and do not let good fortune lift you up: for these things destroy faith.   CLARE OF ASSISI

Live in faith and hope, though it be in darknes, for in this darkness God protects the soul. Cast your care upon God for you are His and He will not forget you. Do not think that He is leaving you alone, for that would be to wrong him.   JOHN OF THE CROSS

Hope always draws the soul from the beauty that is seen to what is beyond, always kindles the desire for the hidden through what is perceived.

GREGORY OF NYSSA

Because the martyrs were devout men and women, fire, flame, wheel and sword seemed to be flowers and perfume to them. If devotion can sweeten the most cruel torments and even death itself, what must it do for virtuous actions?   FRANCIS DE SALES

The disbelief of Thomas has done more for our faith than the faith of the other disciples. As he touches Christ and is won over to belief, every doubt is cast aside and our faith is strengthened. So the disciple who doubted, then felt Christ's wounds, becomes a witness to the reality of the resurrec-tion.   GREGORY THE GREAT

What made the holy apostles and martyrs endure fierce agony and bitter torments, except faith, and especially faith in the resurrection?

FIDELIS OF SIGMARINGEN

THE PRESENCE OF GOD

The early saints describe with unnerving clarity the impossibility of hiding from God or escaping his presence. Later, the emphasis appears to be more on the supportive and inspiring quality of omnipresence, especially in the struggle against sin.

Go in where thou wilt, He sees thee; light thy lamp, He sees thee; quench its light, He sees thee. Fear Him who ever beholds thee. If thou wilt sin, seek a place where He cannot see thee, and then do what thou wilt.

AUGUSTINE OF HIPPO

We avoid the eyes of men, and in God's presence we commit sin. We know God to be the Judge of all, and yet in His sight we sin. AMBROSE

He is never absent, and yet He is far from the thoughts of the wicked; yet He is not absent when far away, for where He is not present by Grace He is present by vengeance. GREGORY THE GREAT

Surely if we remembered that God sees us when we sin, we should never do what displeases Him. JEROME

The prodigal went away and fled into a distant country, but he did not escape from his witnesses, from the accusing eyes of his father.

JOHN CHRYSOSTOM

Who shall dare, in the presence of his prince, to do what displeases that prince? BASIL THE GREAT

In thy strife with the devil thou hast for spectators the Angels and the Lord of Angels. EPHRAEM OF NISIBIS

It behooves thee to be very careful. for thou livest under the eyes of the Judge who beholds all things. In all our thoughts and actions we ought to remember the presence of God, and account all lost in which we think not of Him.   BERNARD OF CLAIRVAUX

This cleric recalls the words of the Apostle: "In God we live and move and have our being." And so he considers himself to be like a fish in the water or a bird in the air, and thus he is always in the presence of God, whom he fears as a Lord who watches him, whom he loves as a Father who provides him with every good thing, and whom he invokes continually and praises and serves without ceasing.   ANTHONY MARY CLARET

In the midst of our employments we ought to have God present to our minds, in imitation of the holy Angels who when they are sent to attend on us so acquit themselves of the function of this exterior ministry as never to be drawn from their interior attention to God.   BONAVENTURE

Be assured that he who shall always walk faithfully in God's presence, always ready to give Him an account of all his actions, shall never be separated from Him by consenting to sin.   THOMAS AQUINAS

Granting that we are always in the presence of God, yet it seems to me that those who pray are in His presence in a very different sense; for they, as it were, see that He is looking upon them, while others may be for days together without even once recollecting that God sees them.

TERESA OF AVILA

He who desires to make any progress in the service of God must begin every day of his life with new ardor, must keep himself in the presence of God as much as possible, and must have no other view or end in all his actions but the divine honor.   CHARLES BORROMEO

He who remembers the presence of God is less open to other thoughts, especially bad thoughts. As long as we believe that God sees us, we are restrained from daring to sin before such a Witness and Judge. In two ways the presence of God is an antidote against sin: first, because God sees us, and, secondly, because we see God.   IGNATIUS OF LOYOLA

CONFORMITY TO
THE WILL OF GOD

Achieving this virtue can bring great inner peace via the route of fatalistic acceptance. There is more of an element of passive suffering in it than in the ideas of trust and reliance on Providence which can create a mood to support daring, if extreme, enterprises. "Not as I will, but as thou wilt" (Matthew 26:39) is likely to be a preparation for sacrifice.

Our whole perfection consists in loving our most amiable God; and all the perfection of the love of God consists in uniting our will with His most holy will. The greatest glory we can give to God is to fulfill His Blessed Will in all things.   ALPHONSUS LIGUORI

If I want only pure water, what does it matter to me whether it be brought in a vase of gold or of glass? What is it to me whether the will of God be presented to me in tribulation or consolation, since I desire and seek only the Divine Will?   FRANCIS DE SALES

The accomplishment of the divine will is the sole end for which we are in the world.   JEAN EUDES

To do the will of God man must despise his own: the more he dies to himself, the more he will live to God.   PETER CLAVER

We must remember that it is God's will, and not our own will, that we must do, for he that doeth His will shall abide forever, even as He abideth forever.   BEDE THE VENERABLE

Entire conformity and resignation to the divine will is truly a road on which we cannot go wrong, and it is the only road that leads us to taste and enjoy that peace which sensual and earthly men know nothing of.   PHILIP NERI

The pleasure which a man seeks in the gratification of his own inclinations is quickly changed into bitterness, and leaves nothing behind but the regret of having been ignorant of the secret of true beatitude and of the way of the saints.   ISIDORE OF SEVILLE

As in heaven Thy will is punctually performed, so may it be done on earth by all creatures, particularly in me and by me.   ELIZABETH OF HUNGARY

Thou alone knowest best and what is for my good. As I am not my own but altogether Thine, so neither do I desire that my will be done, but Thine, nor will I have any will but Thine.   FRANCIS BORGIA

Feed upon the will of God and drink the chalice of Jesus with your eyes shut, so that you may not see what is inside.   PAUL OF THE CROSS

For some time past I had indulged the fancy of offering myself up to the Child Jesus as a plaything, for him to do what he liked with me. I don't mean an expensive plaything; give a child an expensive toy and he will sit looking at it without daring to touch it. But a toy of no value—a ball, say— is all at his disposal; he can throw it on the ground, kick it about, make a hole in it, leave it lying in a corner, or press it to his heart if he feels that way about it.   THÉRÈSE DE LISIEUX

In order to go to God with your heart, your mind must be undisturbed, indifferent. Keep it quiet. Do things simply, without too much analysis. If you really want to please God and intend to be in full agreement with His Will, you can't go wrong.   BL. FRANCIS LIBERMANN

TRUST IN
DIVINE PROVIDENCE

Perfect trust demands a most lively and literal sense of God's reality, of his power, and of his benevolence. Of course it is obvious that Perfect Wisdom can do much better for us than can our own common sense, but it is easy in practice to forget this principle and to try to be clever.

The most practical aspect of trust in Providence, for the saints, is tied to their commitment to poverty. If one relinquishes all material prudence, yet still undertakes great works, then God must provide everything necessary, at just the right moment. The annals of sainthood, ancient and modern, are full of accounts of everyone sitting down to dinner around an empty pot, offering thanks, then answering a knock at the door to find a donor with an extra goose, or the equivalent, in hand. On a larger scale, the great founders and builders habitually sign contracts for thousands or millions, with no cash and no credit, and no anxiety either.

The best natural model of this virtue is the trust of infancy, which we once blissfully knew, before the discovery that human adults are not omnipotent or necessarily benevolent. "Anyone who does not welcome the kingdom of God like a little child will never enter it" (Mark 20:15).

Father, I abandon myself into Your hands; do with me what You will. Whatever You may do, I thank you: I am ready for all, I accept all. Let only Your will be done in me and in all Your creatures—I wish no more than this, Lord.   VEN. CHARLES DE FOUCAULD

I was only an adorer too, of the mystery of the Church, the only ark in the world—and all the heathens savages, sects, etc., were only in my heart for prayer, but never in my brain for what became of them, or to trouble my faith in his wisdom and mercy. The Father most tender, Father of All, my immense God—I his atom.   ELIZABETH SETON

I saw that He is everything that is good and comfortable for us. He is our clothing that for love wrappeth us, claspeth us, and all becloseth us for tender love that He may never leave us.   BL. JULIAN OF NORWICH

I can love you more than you can love yourself and I watch over you a thousand times more carefully than you can watch over yourself. The more trustfully you give yourself up to Me, the more I shall be watching over you; you will gain a clearer knowledge of Me and experience My love more and more joyfully.   CATHERINE OF SIENA

I will not mistrust him, Meg, though I shall feel myself weakening and on the verge of being overcome with fear. I shall remember how Saint Peter at a blast of wind began to sink because of his lack of faith, and I shall do as he did: call upon Christ and pray to him for help. And then I trust he shall place his holy hand on me and in the stormy seas hold me up from drowning.   THOMAS MORE

Tomorrow morning we will agree on some good course to take, under the guidance of the Lord, which He will know how to make redound to the greater glory of his name. For night brings its counsels, as we have often learned from experience.   HUGH OF LINCOLN

I work here on borrowed money, a prisoner for the sake of Jesus Christ. Often my debts are so pressing that I dare not leave the house for fear of being seized by my creditors. When I see so many poor brothers and neighbors of mine suffering, pressed beyond their strength and overwhelmed with physical and mental ills which I cannot relieve, then I become very sorrowful; but I trust in Christ, who knows my heart. Therefore I say, "Woe to the man who trusts in men rather than in Christ."   JOHN OF GOD

The only way to make rapid progress along the path of divine love is to remain very little and to put all our trust in Almighty God. That is what I have done.   THÉRÈSE DE LISIEUX

The exercise of continual abandonment of one's self to the hands of God includes in the most excellent manner all other exercises in their greatest simplicity, purity, and perfection.   FRANCIS DE SALES

The heart of God invites all to put it to the proof. The more He gives, the more He desires to give. He loves to see the trust which makes us persist in knocking unceasingly.   BL. PLACID RICCARDI

When tempted to despair, I have only one resource: to throw myself at the foot of the Tabernacle like a little dog at the foot of his master.
          JOHN VIANNEY

Sometimes we are unduly excited when things go well, and at other times we are too alarmed when things go badly. . . . We ought to establish our hearts firmly in God's strength and struggle, as best we can, to place all of our hope and confidence in the Lord so that we shall be like him, as far as it is possible, even in his unchanging rest and stability.

BL. JORDAN OF SAXONY

In building, we need not act as the people of the world do. They first procure the money and then begin to build, but we must do just the opposite. We will begin to build and then expect to receive what is necessary from Divine Providence. The Lord God will not be outdone in generosity.

ALPHONSUS LIGUORI

We live in troublous times and it seems madness to found a new religious congregation at the very moment that the world and hell together are doing all in their power to wipe out those that exist already. But be not afraid. It is not probabilities but certainties that I am telling you. God is blessing our endeavors and desires them to continue.

JOHN BOSCO

Since our house is open to all, it receives the sick of every kind and condition; the crippled, the disabled, the lepers, mutes, the insane, paralytics, those suffering from scurvy, and those bearing the infirmities of old age, many children, and countless pilgrims and travelers whom we give fire, water, salt, and cooking utensils. We ask no payment for this from anyone, and yet Christ provides for all.

JOHN OF GOD

The more you abandon to God the care of all temporal things, the more He will take care to provide for all your wants; but if, on the contrary, you try to supply all your own needs, Providence will allow you to continue to do just that, and then it may very well happen that even necessities will be lacking, God thus reproving you for your want of faith and reliance in Him.

JEAN BAPTISTE DE LA SALLE

God in his unspeakable providence has arranged that some received the holy reward of their toils even before they set to work, others while actually working, others again when the work was done, and still others at the time of their death. Let the reader ask himself which one of them was made more humble.

JOHN CLIMACUS

There is no such thing as bad weather. All weather is good because it is God's.

TERESA OF AVILA

Now, is not God able to send, perhaps tomorrow, sacks of money to my door?

CAMILLUS DE LELLIS

I have started houses with no more than the price of a loaf of bread and prayers, for with Him who comforts me, I can do anything.

FRANCES XAVIER CABRINI

I place all my confidence in Divine Providence to give my poor boys whatever they need. If, for example, Divine Providence should inspire you, this very moment, to make a generous donation, I would be very grateful, and so would they.

JOHN BOSCO

Cast yourself into the arms of God and be very sure that if He wants anything of you, He will fit you for the work and give you strength.

PHILIP NERI

I worry until midnight and from then on I let God worry.

BL. LOUIS GUANELLA

PRUDENCE, PATIENCE, PERSEVERANCE

The theological virtues of faith, hope, and charity all have God as their immediate objects. Prudence, patience, and perseverance fall in the category of moral virtues that dispose us to lead good lives of right action in the world.

Prudence involves correct judgement and is acquired by one's own acts and learned through trial and error as well as by the infusion of grace. It is the intellectual virtue that allows us to recognize the good and evil possibilities in any situation and to act correctly on the basis of this foresight.

Patience is the moral virtue that enables us to bear hardship and suffering without sorrow or resentment. It is closely related to fortitude, and a bulwark against the sin of anger. Patience is a source of joy for the saints because it is a prime means of strengthening love of God and reaching perfection. It accepts trials as reflections of God's will, making them meaningful and even precious.

Perseverance is the great gift that allows a human being to remain in the state of grace until the end of life. There always remains the possibility of a fall, but trust in Christ's promise that "Anything you ask for from the Father he will grant in my name" (John 16:24) means that with earnest prayer the course can be finished.

Prudence must precede every action that we undertake; for, if prudence be wanting, there is nothing, however good it may seem, which is not turned into evil. **BASIL THE GREAT**

Take care, then, not consciously to do or say anything which, if all the world were to know it, you could not acknowledge and say, "Yes, that was what I did or that was what I said." **LOUIS IX OF FRANCE**

Human, carnal, or worldly prudence is that which has only worldly prosperity in view and is indifferent about the means, provided it attains its object. Christian prudence takes Eternal Incarnate Wisdom for its guide in every thought, word, and work. It is regulated in every emergency not by fatuous, glimmering light of its own, or by worldly judgement, but by the maxims of faith.   VINCENT DE PAUL

Be ye prudent as the serpent who, in danger, exposes his whole body to preserve his head. In the same way, we must risk everything, if necessary, to preserve the love and presence of Our Lord whole and entire within ourselves, for He is our Head and we are His members. FRANCIS DE SALES

Let us study, then, how to bear worldly tribulations with patience, yea, even with cheerfulness, because herein lieth the sign that the Beloved de- lighteth in us and hath chosen us and will give us the pledge of His inheritance.   BL. ANGELA OF FOLIGNO

St. Francis de Sales, that great saint, would leave off writing with the letter of a word half-formed in order to reply to an interruption.   JOHN VIANNEY

To be criticized, denounced and despised by good men, by our own friends and relatives is a severe test of virtue. I admire the patience with which the great St. Charles Borromeo endured the public criticisms which a famous and strictly virtuous preacher directed against him more than his tolerance of all attacks from others.   FRANCIS DE SALES

Let your understanding strengthen your patience. In serenity look forward to the joy that follows sadness.   PETER DAMIAN

There are some favors that the Almighty does not grant either the first, or the second, or the third time you ask Him, because He wishes you to pray for a long time and often He wills this delay to keep you in a state of humility and self-contempt and to make you realize the value of His graces.
                                                              JEAN EUDES

The woman who stayed behind to seek Christ was the only one to see him. For perseverance is essential to any good deed, as the voice of truth tells us: "Whoever perseveres to the end will be saved."   GREGORY THE GREAT

Our body is not made of iron. Our strength is not that of stone. Live and hope in the Lord, and let your service be according to reason. Modify your holocaust with the salt of prudence.   CLARE OF ASSISI

Anthony entreated the vision that appeared, saying, "Where were you? Why didn't you appear in the beginning, so that you could stop my distresses?" And a voice came to him: "I was here, Anthony, but I waited to watch your struggle. And now, since you persevered and were not defeated, I will be your helper forever, and I will make you famous everywhere."
                                          ATHANASIUS, *The Life of Anthony*

Restraining my impatience cost me so much that I was bathed in perspiration.   THÉSÈSE DE LISIEUX

In my long experience very often I had to be convinced of this great truth; that it is easier to become angry than to restrain oneself and easier to threaten a boy than to persuade him. Yes, it is more fitting to be persistent in punishing our own impatience and pride than to correct the boys. We must be firm but kind, and be patient with them.   JOHN BOSCO

SUFFERING

From a natural, worldly point of view, the desire for suffering and the ability to welcome it are probably the most incomprehensible things the saints try to tell us about. We may understand more about sin than we care to admit, and the saints' flights of ecstatic or neighborly love are easy to admire. It is consoling to speculate about Heaven, and the prayer life of other people is always interesting and often inspiring. The saints' craving for suffering, however, may well baffle and embarrass us.

The capacity to embrace suffering is clearly an important and even a central attribute of sainthood. On the natural level, it can be seen as a striving for mastery on the part of supremely realistic people who recognize that suffering is inevitable in human life and must somehow be accepted and transcended before meaningful life is possible. Christian saints must see God's descent into suffering flesh as pointing a way to follow. As Blessed Placid Riccardi says, the Redeemer has "sanctified pain." In terms of the supernatural economy, to be a saint is to be utterly convinced that all suffering that is dedicated to God creates merit and will be rewarded a thousandfold.

Christ tells us that if we want to join him, we shall travel the way he took. It is surely not right that the Son of God should go his way on the path of shame while the sons of men walk the way of worldly honor.

JOHN OF AVILA

Reason should dominate pain, for our Redeemer has sanctified pain and by so doing has given us Christians a right way of facing it. For us, pain does not come to hurt and destroy but to raise to the heights.

BL. PLACID RICCARDI

I shall remind myself of the labors He undertook in preaching, of his weariness while traveling, of the temptations He suffered while fasting, of his vigils while praying, and of the tears He shed out of compassion. I will remember, moreover, his sorrows, and the insults, spittle, blows, ridicule, rebukes, nails, and all the rest that rained down upon Him in abundance.

BERNARD OF CLAIRVAUX

Say always, "My beloved and despised Redeemer, how sweet it is to suffer for you."

ALPHONSUS LIGUORI

Why do you ceaselessly ask to taste of My delights and why do you refuse the tribulations?

MARGARET OF CORTONA

I could not reveal or declare the sweetness I felt or the tears of exceeding great joy that I shed when I was troubled or reviled by my brethren, my friends, or my kindred.

BL. ANGELA OF FOLIGNO

I have had crosses in plenty—more than I could carry, almost. I set myself to ask for the love of crosses—then I was happy.

JOHN VIANNEY

The temple of the spirit is raised through work and suffering; and I would add that suffering counts for more than work.

ANTHONY MARY CLARET

If God is well pleased so long as we do not deny his ordinances, what supreme pleasure we must afford Him when we accept His will with cheerfulness in sufferings that touch our own person. . . . Afflictions and the cauterization of the flesh burn away the rust of sin and perfect the life of the just.

ANSELM OF CANTERBURY

The soul in the darkness groans under its chains, motionless, helpless, until the spirit is softened, humbled, purified, made so subtle, so simple, that it can, in some way, become one with the spirit of God, in accordance with the extent and degree of the union of love to which mercy wishes to raise it.

JOHN OF THE CROSS

My Good Shepherd, who have shown Your very gentle mercy to us unworthy sinners in various physical pains and sufferings, give grace and strength to me, Your little lamb, that in no tribulation or anguish or pain may I turn away from you.

FRANCIS OF ASSISI

Many people would be ready to accept suffering so long as they were not inconvenienced by it. "I wouldn't be bothered by poverty," says one, "If it didn't keep me from helping my friends, educating my children, and living respectably." "It wouldn't bother me," says another, "So long as people didn't think it was my own fault." Or another would be willing to suffer evil lies told about him as long as no one believed his detractors.

FRANCIS DE SALES

We put ourselves to all sorts of inconveniences to satisfy our guilty passions but when it is a question of overcoming them we will not lift a finger. It is just this penny's worth of suffering that nobody wants to spend.

LEONARD OF PORT MAURICE

An unpitied pain wins greater merit before God. Never say to God: "Enough"; simply say, "I am ready!"   BL. SEBASTIAN VALFRE

Try to resemble that holy soul who wouldn't give up the least part of her sufferings so as not to lose the merit thereof.   JOSEPH CAFASSO

Suffering is a short pain and a long joy.   BL. HENRY SUSO

May it please the Mother of God to hear my prayer for you, which is that you may meet with even greater affronts, so that you may have the occasion of greater merit, provided that you can accept them with patience and consistency and without sin on the part of others, remembering the greater insults which Christ our Lord suffered for us.   IGNATIUS OF LOYOLA

There'll be a lot of surprises at the Last Judgement when we shall be able to see what really happened inside people's souls; and I think the way of suffering by which God led me will be a revelation to the people who knew me.   THÉRÈSE DE LISIEUX

When it is all over, you will not regret having suffered; rather you will regret having suffered so little and suffered that little so badly.

BL. SEBASTIAN VALFRE

To suffer much, yet badly, is to suffer like reprobates. To suffer much, even bravely, but for a wicked cause, is to suffer as a martyr of the devil. To suffer much or little for the sake of God is to suffer like saints.

LOUIS-MARIE GRIGNION DE MONTFORT

What I suffered is known only to One for whose love and in whose cause it is pleasing and glorious to suffer.   ISAAC JOGUES

I would willingly endure all the sufferings of this world to be raised a degree higher in Heaven, and to possess the smallest increase of the knowledge of God's greatness.   TERESA OF AVILA

O dignity, O holiness, O admirable excellence of suffering used for the perfection and consummation of a God, for the perfection of Jesus, God and Man, for the consummation of Him Who is the consummation and perfection of all things! Great humiliation of Jesus, Who humbled Himself to a state in which He was capable of being perfected and consummated! And the immense dignity of suffering chosen and used by Him and His eternal Father to achieve this perfection and consummation!   JEAN EUDES

TRIALS

Trials may consist of a host of natural and divinely sent sources of suffering that give the believer an opportunity to offer God a sacrifice of praise and to draw closer to him by strengthening in virtue, deepening of trust in his justice, and acceptance of his will. "Blessed is the man whom thou chasteneth, O Lord, and teachest out of thy law" (Psalms 94:12). Job's attitude represents an ideal: "Though he slay me, yet will I trust in him" (Job 13:15).

Temptation, which comes from the world, the flesh, or the devil, may be a special sort of trial, as when Jesus was tested in the desert by Satan. Overcoming temptation strengthens virtue in specific ways. Humility opposes pride, meekness conquers anger, and charity stifles envy. God has promised to support us in this test. "God is faithful who will not suffer you to be tempted above that ye are able; but will with the temptation also make a way to escape that ye may be able to bear it" (I Corinthians 10:13).

Place a nail on a board. Will it ever go through the wood on its own, no matter how sharp it is? No, indeed. You will only sink it into the board by hitting it with a hammer. We are just the same; it is only by hammer blows that God manages to humble us, no matter how good our native dispositions might be.   ANTHONY MARY CLARET

So great are the trials, and so profound the darkness, spiritual as well as corporal, through which souls must pass, if they will attain to perfection, that no human learning can explain them, nor experience describe them. He only who has passed through them can know them, but even he cannot explain them.   JOHN OF THE CROSS

This most holy tribulation is exceedingly profitable for us; wherefore let us not avoid it nor hold it in horror. For of a certainty and with all mine heart say I unto you that our noblest advocates and truest witnesses, those who will be most readily believed in the presence of God, are precisely these holy and precious tribulations, whose worth we know not; for with them do we purchase the kingdom of heaven, and the joys of eternity are obtained through poverty, tears, suffering, and persecution.

BL. ANGELA OF FOLIGNO

There are people who make capital out of everything, even the winter. If it is cold, they offer their little sufferings to God. JOHN VIANNEY

I pray God may open your eyes and let you see what hidden treasures he bestows on us in the trials from which the world thinks only to flee.

JOHN OF AVILA

God's grace has given the king a gracious frame of mind toward me, so that as yet he has taken from me nothing but my liberty. In doing this His Majesty has done me such great good with respect to spiritual profit that I trust that among all the great benefits he has heaped so abundantly upon me I count my imprisonment the very greatest. I cannot, therefore mistrust the grace of God. THOMAS MORE

It would not be difficult for Him to free us at once from all tribulation and sorrow, but instead He permits His friends to suffer much in this world that instead He may crown them all the more gloriously in Heaven, and make them more like his only begotten son, who never ceased to do good and to suffer injury while he was on earth that He might teach us patience by his example. ROBERT BELLARMINE

We are too little to be able always to rise above difficulties. Well, then, let us pass beneath them quite simply. THÉRÈSE DE LISIEUX

They stripped my companion naked, and like mad dogs ripped away his nails with their teeth, bit his fingers off, and pierced his right hand with a spear. He endured it all with invincible patience, filled with the thought of the nails that fastened our Saviour. They turned on me with their fists and knotted sticks, left me half-dead on the ground, and a little later tore away my nails in the same way, and bit off my two forefingers which caused me incredible agony. ISAAC JOGUES

God wishes to test you like gold in the furnace. The dross is consumed by the fire, but the pure gold remains and its value increases. It is in this manner that God acts with his good servant, who puts his hope in him and remains unshaken in times of distress. God raises him up, and in return for what he has given up out of love for God, he is repaid a hundredfold in this life and with eternal life hereafter. JEROME EMILIANI

The way that God deals with men can only be praised. He lashes them in this life to shield them from the eternal lash in the next. He pins peole down now; at a later time he will raise them up. He cuts them before healing; he throws them down to raise them anew.   PETER DAMIAN

Is not the life of man upon earth a trial? Who would want troubles and difficulties? You command us to endure them, not to love them. No person loves what he endures, though he may love the act of enduring.

AUGUSTINE OF HIPPO

SACRIFICE

When love and fear are alive and urgent, the impulse to sacrifice is spontaneous. This impulse is natural to all love, sacred and profane, and is the instinctive way of propitiating the unseen. It stands at the heart of all religious systems.

For the saints, nothing is too great or too trivial to be an acceptable sacrifice: a wooden cup, the human will, one's health, a much loved child, any kind of dull toil or suffering rightly dedicated, the relinquishing of some pleasure or satisfaction, Christ's life for us, our life for him.

When we do something we dislike, let us say to God: "My God, I offer you this in honor of the moment when you died for me." JOHN VIANNEY

One Lent he was whittling a little cup to occupy his spare moments and to prevent them from being wasted. When he was reciting Terce, it came into his mind and distracted him a little. Moved by fervor of spirit, he burned the cup in the fire, saying: "I will sacrifice this to the Lord, whose sacrifice it has impeded." BONAVENTURE, *The Life of St. Francis*

He, the most beatific and most joyful, did make Himself most wretched in order that through His infinite suffering He might redeem man and save him from everlasting and unspeakable pain. BL. ANGELA OF FOLIGNO

I will attempt day by day to break my will into little pieces. I want to do God's holy will, not my own. GABRIEL POSSENTI

In the old Law, God would accept no victim as a holocaust if it had not first been flayed; in like manner, our hearts can never be immolated and sacrificed to God until they shall have been stripped of the old skin of their habits, inclinations, repugnances and superfluous affections.

FRANCIS DE SALES

The Lord's command seems difficult and painful when he says that anyone who wishes to follow him must deny himself. But it is not really so, since he helps us to do what he commands. He fulfills his own words: "My yoke is easy and my burden is light." Love makes it easy to carry out whatever is difficult in his command.   AUGUSTINE OF HIPPO

I was a free man in a worldly position; my father was a decurion. Indeed, I bargained away my aristocratic status—I am neither ashamed nor sorry— for the benefit of others. In short, I am a slave in Christ to an outlandish nation because of the unspeakable glory of eternal life which is in Christ Jesus our Lord.   PATRICK OF IRELAND

We should not attach much value to what we have given God, since we shall receive for the little we have bestowed upon Him much more in this life and in the next.   THERESA OF AVILA

If God is making your son His own, as well as yours, so that he may become even richer, even more noble, even more distinguished and, what is better than all this, so that from being a sinner he may become a saint, what do either you or he lose? You are not losing him. On the contrary, through him you are gaining many sons. All of us at Clairvaux or of Clairvaux will receive him as a brother and you as our parents.

          BERNARD OF CLAIRVAUX

I am God's wheat and I am ground by the teeth of wild beasts that I may be found pure bread of Christ.   IGNATIUS OF ANTIOCH

I thank God that now when I preach I shall be able to say instead of "dear brethren," "my fellow lepers."

          VEN. JOSEPH DE VEUSTER (FATHER DAMIEN OF MOLOKAI)

Christ one day said to St. John of the Cross, "John, what recompense dost thou ask for thy labors?" He answered: "Lord, I ask no other recompense than to suffer and be contemned for Thee."   JOHN OF THE CROSS

I began to understand that the love of the sacred heart without a spirit of sacrifice is but empty illusion.   BL. MARIA DROSTE ZU VISCHERING

I will take no unnecessary walks.
     I will make exactingly careful use of my time.
     When the salvation of souls is at stake I will always be ready to act, to suffer, and to humble myself.
     May the charity and gentleness of St. Francis de Sales inform my every action.

I will always be content with the food set before me unless it is really harmful to my health.

I will always add water to my wine and drink it only for reasons of health.

Since work is a powerful weapon against the enemies of my salvation I will take only five hours sleep a night. During the day, especially after dinner, I will take no rest, except in case of illness.

Every day I will devote some time to meditation and spiritual reading. During the day I will make a short visit, or at least a prayer, to the Blessed Sacrament. My preparation for Mass shall last at least a quarter of an hour and so shall my thanksgiving.

Outside the confessional and save in cases of strict necessity I will never stop to talk to women.   JOHN BOSCO

I do not suffer; at least I suffer without suffering by a sacrifice of not being concerned whether I receive esteem or disesteem, approval or disapproval, contempt or praise.   PETER JULIAN EYMARD

It's a form of trade, you see. I ask God for souls, and pay him by giving up everything else.   JOHN BOSCO

How terrible, I thought, that no act of love is ever made in hell! And I told God that I was ready to go there myself, if it pleased him to contrive, in that way, that for all eternity there would be one loving soul in that abode of blasphemy.   THÉRÈSE DE LISIEUX

THE CROSS

_____

The cross is the most important of all Christian symbols, standing as it does for the instrument on which Christ sacrificed himself to redeem the world. It also stands for the suffering and sacrifice that all followers of Christ must be prepared to accept willingly in order to be joined with him and to share in the salvation of souls. "If any man will come after me, let him deny himself, and take up his cross, and follow me" (Matthew 16:24).

Crucifixion was a particularly shameful method of execution, considered by Jews and Romans alike to be a symbol of servitude to the Roman overlords. No one who could prove Roman citizenship was liable to this penalty. In addition to sacrifice and suffering, then, the cross symbolizes humiliation to an extreme degree, making it utterly clear that Christ's kindgom is not of this world. The cross thus stands for a reversal of natural, reasonable values. "Base things of the world and things which are despised hath God chosen" (I Corinthians 1:27) to demonstrate his transcendent power and purposes.

_____

The Cross possesses such power and strength that, whether they will or no, it attracts, draws, and ravishes those who carry it.   BL. HENRY SUSO

As through a tree we were made debtors to God, so through a tree we receive cancellation of our debt.   IRENAEUS

Consider whether the business of the world could be carried on without the figure of the cross. The sea cannot be crossed unless this sign of victory, the mast, remains upright. Without it there would be no ploughing, nor could diggers or mechanics do their work without their cross-shaped tools. Humans are distinguished from beasts by their upright posture and the way they can extend their arms. The very nose through which we get our breath is set at right angles to the brow, fulfilling the prophet's words, "Our Lord Christ is the breath before our face."   JUSTIN MARTYR

If you would understand that the cross is Christ's triumph, hear what he himself also said: "When I am lifted up, then I will draw all men to myself." Now you can see that the cross is Christ's glory and triumph.

ANDREW OF CRETE

The Cross to me is certain salvation. The Cross is that which I ever adore. The Cross of the Lord is with me. The Cross is my refuge.

THOMAS AQUINAS

I shrink every time I look at the cross because I feel I could die thinking of the pain of it, yet, in spite of this deep repugnance of mine, my heart welcomes all the sufferings it entails and in these I find all my delight.

GEMMA GALGANI

These perfect souls as if enamoured of my honor and famished for the food of souls rush to the table of the most holy Cross. They are willing to suffer all sorts of pain and to endure much in service of their neighbor. They wish ardently to persevere and acquire all virtues and to bear in their body the stigmata of Christ crucified. They cause the crucified love which is theirs to shine forth, manifest in their inner joy and their delighted endurance of the shames and vexations that surround them.   ·

CATHERINE OF SIENA

Ye shall therefore have the blessing which I shall give at the last judgement, inasmuch as ye did not repulse Me when I came into Mine own place, as did my persecutors, but of your compassion did receive Me into the lodging of your hearts as a desolate pilgrim; when I hanged naked upon the Cross, hungry, thirsty, and sick, and pierced by nails, ye did suffer with Me in my death and desired in all things to be My companions.

BL. ANGELA OF FOLIGNO

You must accept your cross; if you carry it courageously it will carry you to heaven.   JOHN VIANNEY

Hold out, be steadfast, endure, bear the delay, and you have carried the cross.   AUGUSTINE OF HIPPO

The Christian who desires to follow Jesus carrying his cross must bear in mind that the name "Christian" means "learner or imitator of Christ" and that if he wishes to bear that noble title worthily he must above all do as Christ charges us in the Gospel; we must oppose or deny ourselves, take up the cross, and follow him.   ANTHONY MARY CLARET

Do as the storekeeper does with his merchandise; make a profit on every article. Suffer not the loss of the tiniest fragment of the true cross. It may only be the sting of a fly or the point of a pin that annoys you; it may be the little eccentricities of a neighbor, some unintentional slight, the insignificant loss of a penny, some little restlessness of soul. a slight physical weakness, a light pain in your limbs. Make a profit on every article as the grocer does, and you will soon be wealthy in God.

LOUIS-MARIE GRIGNION DE MONTFORT

I see an infinite number of crucified persons in the world, but few who are crucified by the love of Jesus. Some are crucified by their self-love and inordinate love of the world, but happy are they who are crucified for the love of Jesus; happy are they who live and die on the cross with Jesus.

JEAN EUDES

Of all the pains that lead to salvation this is the most pain, to see thy Love suffer. How might any pain be more to me than to see Him that is all my life, all my bliss, and all my joy suffer? Here I felt soothfastly that I loved Christ so much above myself that there was no pain that might be suffered like to that sorrow that I had to see Him in Pain.   BL. JULIAN OF NORWICH

No one would complain about his cross or about troubles that may happen to him if he could come to know the scales on which they are weighed when they are distributed to men.   ROSE OF LIMA

Whenever anything disagreeable or displeasing happens to you remember Christ crucified and be silent.   JOHN OF THE CROSS

PERSECUTION
AND MARTYRDOM

———

Martyrdom is the most perfect act of a Christian since it is a perfect proof of love, the greatest of all virtues. "Greater love hath no man than this, that a man lay down his life for his friends" (John 15:13). Martyrs imitate the sacrifice of Christ in accepting death on behalf of the world's salvation. The term means a "witness" for the faith who accepts death by refusing to apostatize.

The traditional presumption that genuine martyrdom confers sanctity still holds, as shown in recent canonizations of groups en masse: Paul Miki and the Twenty-five Martyrs of Japan (1862); the Jesuit North American Martyrs (1930); Charles Lwanga of Uganda and his Twenty-one Companions (1964); the Forty Martyrs of England and Wales (1970).

———

Nothing is more to be feared than too long a peace. You are deceived if you think that a Christian can live without persecution. He suffers the greatest persecution of all who lives under none. A storm puts a man on his guard and obliges him to exert his utmost efforts to avoid shipwreck.   JEROME

Repeat to yourself in every danger: Whether we live or whether we die, we are the Lord's. How glorious is the conqueror's return from battle, how blessed are the martyrs who do not return! Rejoice, brave soldier, if you survive and conquer in the Lord, but rejoice and glory still more if you die and are thus joined to the Lord. Life, indeed, is fruitful, and victory a splendid thing, but death by sacred right is to be preferred to both, for if blessed are those who die in the Lord, are they not to be accounted more blessed far who die for the Lord?   BERNARD OF CLAIRVAUX, *to the Templars*

The martyrs are perfected in righteousness, and they earned perfection through their martyrdom. For them the church does not pray: for the other departed faithful she prays, but not for martyrs. They have gone out of this world so perfected that instead of being our clients they are our advocates.   AUGUSTINE OF HIPPO

A man may very well lose his head and yet come to no harm—yea, I say, unspeakable good and everlasting happiness.   THOMAS MORE

If you require my estate, you may take it; if my body, I readily give it up; have you a mind to lead me with irons or put me to death, I am content. I shall not fly to the protection of the people nor cling to the altar; I choose rather to be sacrificed for the sake of the altar.   AMBROSE

I pray, my brother, that we may be found worthy to be cursed, censured, and ground down, and even to be executed in the name of Jesus Christ, as long as Christ Himself is not killed in us.   PAULINUS OF NOLA

I will not purchase corruptible life at so dear a rate; and indeed, if I had a hundred lives, I would willingly lay down all in defense of my faith.
                                                        BL. CHRISTOPHER BUXTON

When you see our heads fixed up over the bridge, think that they are there to preach to you the very same faith for which we are about to die.
                                                                    ALBAN ROE

Be it known unto you that we have made a league—all the Jesuits in the world—cheerfully to carry the cross that you shall lay upon us and never to despair your recovery while we have a man left to enjoy your Tyburn, or to be racked with your torments, or to be consumed with your prisons. The expense is reckoned, the enterprise is begun; it is of God—it cannot be withstood. So the Faith was planted, so it must be restored.
                                                              EDMUND CAMPION

There are many among the martyrs of my age or younger, and as weak or weaker than I; but the Divine Grace that did not fail them will sustain me.
                                                              ROBERT SOUTHWELL

Calling to mind the sacrifices I had offered thee in Thy Church, I took hold of the amputated thumb with my other hand and offered it to Thee, my God, living and true, until on the advice of a companion I let the thumb drop, for fear they should put it in my mouth and force me to swallow it as they often do. I thank God they left me the one on my right hand and so enabled me by this letter to beg my Fathers and brethren to offer petition for us in the Holy Church of God, whose custom it is to pray "for the afflicted and for captives."   ISAAC JOGUES

I pray to God to fortify the martyrs, but I do not ask Him to preserve them.   BL. MARY AMADINE OF CHINA

Why should others cause me to offend God, or to lose the charity which I owe and bear them? If any person were to cut off my arms or pluck out my eyes, they would be the dearer to me, and would seem the more to deserve my tenderness and compassion.   EDMUND THE MARTYR

If you want to join me in China, hurry, for I will soon have my head cut off for Jesus.   BL. MARIA CHIARA OF CHINA

Now I myself suffer what I am suffering; but then there will be Another in me who will suffer for me, because I am to suffer for him.   FELICITY OF CARTHAGE

For the sake of my God whom I love, I ask him that he may grant that I may spill my blood along with those other exiles and prisoners even though I may lack burial itself or my corpse may be most squalidly torn limb from limb by dogs or wild beasts or the birds of the air may devour it. I believe most confidently that if this were to happen to me I have gained my soul along with my body.   PATRICK OF IRELAND

I can't describe the pleasure, delight, and joy I felt in my soul on realizing that I had reached the long-desired goal of shedding my blood for the love of Jesus and Mary and of sealing the truths of the Gospel with the very blood of my veins. My spirits soared at the thought that this was but a promise of what I might achieve in time—to shed my blood completely in the ultimate sacrifice of death.   ANTHONY MARY CLARET

Now at last I begin to be a disciple! Let nothing visible or invisible hinder me, through jealousy, from attaining to Jesus Christ. Come fire, come cross, come whole herds of wild beasts, come drawing and quartering, scattering of bones, cutting off of limbs, crushing of the whole body, all the horrible blows of the devil—let all these things come upon me, if only I may be with Christ.   IGNATIUS OF ANTIOCH

We hope to suffer torment for the sake of our Lord Jesus Christ and so be saved.   JUSTIN MARTYR

In the midst of tears, she sheds no tears herself. She stood still, she prayed, she offered her neck. You could see fear in the eyes of the executioner, as if he were the one condemned. His right hand trembled, his face grew pale as he saw the girl's peril, while she had no fear for herself.   AMBROSE, *on the death of Saint Agnes*

As I come to this supreme moment of my life, I am sure none of you would suppose I want to deceive you. And so I tell you plainly: there is no way to be saved except the Christian way. My religion teaches me to pardon my enemies and all who have offended me. I do gladly pardon the Emperor and all who have sought my death. I beg them to seek baptism and be Christians themselves.   PAUL MIKI

Leave me as I am. The one who gives me strength to endure the fire will also give me strength to stay quite still on the pyre, even without the precaution of your nails.   POLYCARP

The crowd in wonder watched God's heavenly contest, this spiritual battle that was Christ's. They saw his servants standing firm, free in speech, undefiled in heart, endowed with supernatural courage, naked and deprived of the weapons of this world, but as believers equipped with the arms of faith. Tortured men were stronger than their torturers; battered and lacerated limbs triumphed over clubs and claws that tore them.

                                        CYPRIAN OF CARTHAGE

A fountain fed from many springs will never dry up. When we are gone, others will rise in our place.   BRUNO SERUNKUMA OF UGANDA

I know not your gods. Jesus Christ, the only Son of God is my God. Beat, tear, or burn me, and if my words offend you, cut out my tongue; every part of my body is ready when God calls for it as a sacrifice.

                                        THEODORE OF HERACLEA

When the holy martyrs were led into Alexandria, Anthony left his cell and followed, saying "Let us go also, that we may enter the combat, or look upon those who do." He yearned for martyrdom, but because he did not wish to hand himself over, he served the confessors in the mines and in the prisons. In the law court, he showed great enthusiasm, stirring to readiness those who were called forth as contestants, receiving them as they underwent martyrdom, and remaining in their company until they were perfected.   ATHANASIUS, *Life of Anthony*

The martyrs desired death, not to fly labor, but to attain their end. And why did they not fear death, from which man so naturally shrinks? Because they had vanquished the natural love of their own bodies, by divine and supernatural love.   CATHERINE OF SIENA

God does not require of us the martyrdom of the body; he requires only the martyrdom of the heart and the will.   JOHN VIANNEY

We should not worry about equality, but I do believe that the martyrdom of love cannot be relegated to second place, for "love is as strong as death."

The martyrs of love suffer infinitely more in remaining in this life so as to serve God, than if they died a thousand times over in testimony to their faith, their love, and their fidelity.   JEANNE FRANÇOISE DE CHANTEL

Our friends, then, are all those who unjustly afflict upon us trials and ordeals, shame and injustice, sorrows and torments, martyrdom and death; we must love them greatly for we will possess eternal life because of what they bring upon us.   FRANCIS OF ASSISI

Christianity is a warfare, and Christians spiritual soldiers. In its beginning, our faith was planted in the poverty, infamy, persecution and death of Christ; in its progress, it was watered by the blood of God's saints; and it cannot come to the full growth unless it be fostered with the showers of the martyrs' blood.   ROBERT SOUTHWELL

THE EUCHARIST

When in the Mass the priest pronounces the words of eucharistic consecration, "This is my body, this is my blood," the underlying reality of the bread and wine is changed into the body and blood of Christ, given for us in sacrifice. Christ himself is really and substantially present, in a mysterious way, under the appearances of bread and wine.

Eucharist means "thanksgiving" because Christ gave thanks at the Last Supper before he broke the bread. It refers to the miraculous real presence of Christ in the bread and wine of the Mass, to the sacramental sacrifice of the Mass, and to the communion of the faithful whereby Christ nourishes the soul. "He who eats my flesh and drinks my blood has life everlasting, and I will raise him up on the last day" (John 6:54).

The consecrated host is reserved in the church or chapel for personal devotions. At specific times it may be exposed for veneration, as in the Forty Hours Devotion or at the Benediction of the Blessed Sacrament. There are a number of religious orders and even secular parishes devoted to perpetual adoration of the eucharistic host, where the members take their turn in around-the-clock prayer and contemplation in the presence of the exposed Sacrament.

This morning my soul is greater than the world since it possesses You, You Whom heaven and earth do not contain.   MARGARET OF CORTONA

This heavenly bread demands that the interior man hunger for it, and it satiates such only as render themselves worthy of it by the fervor of their desires.   AUGUSTINE OF HIPPO

When you place a Eucharistic spark in a soul, you have implanted therein a divine seed of life and of all the virtues which is self-sufficient.

PETER JULIAN EYMARD

Our Savior has instituted the most august sacrament of the Eucharist, which contains his flesh and blood in their reality, so that whoever eats of it shall live forever. Therefore, whoever turns to it frequently and devoutly so effectively builds up his soul's health that it is almost impossible for him to be poisoned by evil affection of any kind. FRANCIS DE SALES

Yesterday, on approaching the Most Blessed Sacrament, I felt myself burning and I had to withdraw. I am astounded that so many who receive Jesus are not reduced to ashes. GEMMA GALGANI

When I am before the Blessed Sacrament I feel such a lively faith that I can't describe it. Christ in the Eucharist is almost tangible to me. I kiss his wounds continually and embrace Him. When it's time for me to leave, I have to tear myself away from his sacred presence. ANTHONY MARY CLARET

I have many times seen the Body of Christ in divers forms in this Blessed Sacrament. For sometimes I have seen the throat of Christ more splendid and beauteous than the sun, and by that beauty was it certified to me that God Himself was here, seeing that it was incomparably greater than the sun both in beauty and quantity, wherefore it doth greatly grieve me that I cannot make it manifest. Sometimes I have seen two eyes of great splendour, and so large that I beheld nothing of the Host save the edge thereof. BL. ANGELA OF FOLIGNO

It is easy for me to distinguish between the body of Christ and a simple wafer of unleavened bread. If this host had been consecrated I should have swallowed it without the least difficulty; but I know that this one has not been for my whole nature revolts against consuming it. BL. LYDWINA OF SCHIEDAM

As the bread, which comes from the earth, on receiving the invocation of God, is no longer common bread but Eucharist, and is then both earthly and heavenly; so our bodies, after partaking of the Eucharist, are no longer corruptible, having the hope of the eternal resurrection. IRENAEUS

Surely such a stroke of love should be enough to awaken anyone from the slumber of indifference. Let such a one reflect upon this Mystery and say to himself: "It is God Almighty who will come down upon the altar at the words of consecration; I shall hold him in my hands, and converse with Him, and receive Him into my breast." JOHN OF AVILA

As a woman, compelled by natural affection, hastens to feed her babe from her overflowing breast, so also Christ ever nourishes with His Blood those whom He regenerates. JOHN CHRYSOSTOM

The mother may give her child suck of her milk, but our precious Mother, Jesus, He may feed us with Himself, and doeth it full courteously and full tenderly, with the Blessed Sacrament that is precious food of very life.

BL. JULIAN OF NORWICH

If as often as the Lord's blood is shed, it is poured forth for the remission of sins, I ought to receive it always, so that my sins may always be forgiven. I who am always committing sin ought always to have a remedy.

AMBROSE

Our Lord does not come from Heaven every day to stay in a golden ciborium. He comes to find another Heaven, the Heaven of our soul in which He loves to dwell.   THÉRÈSE DE LISIEUX

Many mothers there are who after the pains of childbirth give their children to strangers to nurse. But Christ could not endure that His children should be fed by others. He nourishes us Himself with his own blood and in all ways makes us one with Himself.   JOHN CHRYSOSTOM

Jesus Christ found a way by which he could ascend into Heaven and yet remain on the earth. He instituted the adorable Sacrament of the Eucharist so that he might stay with us and be the Food of our soul; that he might stay with us and be our Companion.   JOHN VIANNEY

The proper effect of the Eucharist is the transformation of man into God.

THOMAS AQUINAS

Let us suppose that the last day has come and that our doctrine of the Eucharist turns out to be false and absurd. If our Lord now asks us reproachfully: "Why did ye believe thus of my Sacrament? Why did ye adore the Host?" may we not safely answer him: "Yea, Lord, if we were wrong in this, it was you who deceived us. We heard your word, *This is my Body*, and was it a crime for us to believe you? We were confirmed in our mistake by a hundred signs and wonders which could have had you only for their author. Your Church with one voice cried out to us that we were right, and in believing as we did we but followed in the footsteps of all your saints and holy ones."   ROBERT BELLARMINE

THE CHURCH

The church is most broadly defined as the faithful of the whole world. She is "the mystical body of Christ" because he is the invisible head and the faithful are the visible members. The body of the church includes those living on earth (the Church Militant), those in purgatory (the Church Suffering), and the saints in heaven (the Church Triumphant).

She is both human and divine, present in this world, and yet not at home in it. Her function is to act as a sign of God's presence in the world and to carry on Christ's work under the guidance of the Holy Spirit.

The Church is also the mystical bride of Christ, "arrayed in fine linen, clean and white: for the fine linen is the righteousness of saints" (Revelation 19:8).

The early fathers, like Irenaeus, Cyprian, and Ambrose, are engaged in defining the nature of the Church and asserting her integrity and necessity as the vessel of salvation, like the ark of Noah adrift in the rough waters of the pagan world.

As she becomes ascendent, the saints turn toward admiration of her parts and functions—the Church Contemplative and the Church of Action in the world. This admiration is at times sorely tried during periods of worldliness, abuses, and corruption.

The church of the Lord is built upon the rock of the apostles among countless dangers in the world; it therefore remains unmoved. The Church's foundation is unshakable and firm against the assaults of the raging sea. Waves lash at the Church but do not shatter it. Although the elements of the this world constantly batter and crash against her, she offers the safest harbor of salvation for all in distress.   AMBROSE

The Church is like a great ship being pounded by the waves of life's different stresses. Our duty is not to abandon ship, but to keep her on her course.   BONIFACE

Where the Spirit of God is, there is the Church and every kind of grace. The Spirit is truth. Therefore whose who have no share in the Spirit are not nourished and given life at their mother's breast, nor do they enjoy the sparkling fountain that issues from the body of Christ.　　IRENAEUS

By the word of His power He gathered us out of all lands, from one end of the earth to the other end of the world, and made resurrection of our minds, and remission of our sins, and taught us that we are members one of another.　　ANTHONY OF EGYPT

The Church, which is founded upon Christ, received from him the keys of the kingdom of Heaven, that is, the power of binding and forgiving sins, in the person of Peter. Therefore this church, by loving and following Christ, is set free from evil.　　AUGUSTINE OF HIPPO

He cannot have God for his father who has not the Church for his mother. If anyone was able to escape outside of Noah's ark, then he also escapes who is outside the doors of the Church.　　CYPRIAN

Consider everything carefully, and you will find that the customs of the Holy Roman Church do not differ at all from the divine authorities.
　　ISIDORE OF SEVILLE

O Truly blessed mother Church, whom the glorious blood of the victorious martyrs adorns, whom the white garment of virginity clothes with an inviolate confession of praise, neither roses nor lilies are wanting to thy garments!　　BEDE THE VENERABLE

Remember then how our fathers worked out their salvation; remember the sufferings through which the Church has grown, and the storms the ship of Peter has weathered because it has Christ on board.　　THOMAS BECKET

With all the sweet Sacraments He sustaineth us full mercifully and graciously. And so meant He in this blessed word where that He said: "I it am that Holy Church preacheth thee and teacheth thee," That is to say: "All the health and life of Sacraments, all the virtue and grace of my word, all the Goodness that is ordained in Holy Church for thee, I it am."　　BL. JULIAN OF NORWICH

To be with the Church of Jesus Christ with but one mind and one spirit, we must carry our confidence in her, and our distrust of ourselves, so far as to pronounce that true which appears to us false, if she decides that it is so; for we must believe without hesitation that the Spirit of our Lord Jesus Christ is the spirit of his spouse, and that the God who formerly gave the decalogue is the same God who now inspires and directs his church.　　IGNATIUS OF LOYOLA

May no temporal prince presume by any law to take upon him, as rightfully belonging to the see of Rome, a spiritual preeminence by the mouth of our Savior himself, personally present upon the earth, to Saint Peter and his successors.   THOMAS MORE

Now it seems that Highest and Eternal Goodness is having that done by force which has not been done willingly; it seems that He is permitting dignities and luxuries to be taken away from His Bride, as if He would show that Holy Church should return to her first condition, poor, humble and meek as she was in that holy time when men took note of nothing but the honor of God and the salvation of souls, caring for spiritual things and not for temporal. For since she has aimed more at temporal than at spiritual, things have gone from bad to worse.

CATHERINE OF SIENA, *to Pope Gregory XI*

The passable and temporal life of Jesus in His mystical body, that is, in all Christians, has not yet reached its accomplishment, but develops itself from day to day in each true Christian and will not be perfectly complete until the end of time.   JEAN EUDES

The aim of the apostolate is the growth of the Mystical Body through the conversion of sinners, the salvation of souls throughout the world, and the perseverance and growth of all the just in Christ. This is the same goal that drew down the Son of God to earth.   ANTHONY MARY CLARET

Christianity is not dead, as the gentlemen of the Voltaire school like to think. I know some young men living in the world, in the very center of the greatest riches and luxury, yet humble, good, devout, charitable, reverent, seeking out the poor in their garrets, "religious as a woman," as the saying is. Their manners are simple and natural for they are thoroughly in earnest . . . All their lives are spent in doing good.

BL. THEOPHANE VENARD

RELIGIOUS LIFE

Religious life is lived in accordance with the three public vows of poverty, chastity, and obedience. There are many varieties of the professional religious life, from the solitary hermitage, through enclosed or cloistered communities, to orders that exist entirely "in the world."

Anthony of Egypt laid the foundations of monasticism when he withdrew to the desert to live a hermit's life and was followed by other who created a model for the City of God, which so thrilled Athanasius and John Chrysostom.

Most comments by "insiders" on the communal religious life acknowledge that it presents a microcosm of the trials offered by the world, albeit in a more manageable form, as well as some temptations unique to itself. As such, it is an ideal training ground for increase of virtues by means of trial, struggle, and triumph.

In the Church there are two kinds of life, the active and the contemplative. The active life is lower and the contemplative life is higher. The active life is such that it begins and ends on earth. The contemplative life, however, may indeed begin on earth, but it will continue without end into eternity. This is because the contemplative life is Mary's part which shall never be taken away. The active life is troubled and busy about many things, but the contemplative life sits in peace with the one thing necessary.

ANONYMOUS, *The Cloud of Unknowing*

All monastic life may be said to take one of three forms. There is the road of withdrawal and solitude for the spiritual athlete; there is the life of stillness shared with one or two others; there is the practice of living patiently in community.   JOHN CLIMACUS

So their cells in the hills were like tents filled with divine choirs—people chanting, studying, fasting, praying, rejoicing in the hope of future good, working for the distribution of alms, and maintaining both love and harmony among themselves. It was as if one truly looked on a land all its own—a land of devotion and righteousness.

ATHANASIUS, *The Life of Anthony*

Come and see the tents of the soldiers of Christ; come and see their order of battle; they fight every day, and every day defeat and immolate the passions which assail us. JOHN CHRYSOSTOM

Come and behold here a noble youth brought up to inherit great wealth and of noble lineage who, forsaking mother and brothers and sisters, went forth to live on an island, in a rough, deserted spot, alone in a solitude where nothing is heard save the roar of the ocean. There he stands alone. But no, I cannot say alone, for he is in the company of Jesus Christ.

JEROME

Dear cell, in which I have spent such happy hours, with the wind whistling through the loose stones and the sea spray hanging on my hair! COLUMBA

Tedium is rebuffed by community life, but she is a constant companion of the hermit, living with him until the day of his death, struggling with him until the very end. She smiles at the sight of a hermit's cell and comes creeping up to live nearby. JOHN CLIMACUS

Toiling and seeking the fear of God in patience and quiet, they achieve the true manner of life, because their souls are ready to follow the love of God. This is the first kind of calling. ANTHONY OF EGYPT

This is self-renunciation—to unlock the chains of this earthly life which passeth away and to set oneself free from the business of men, and thus to make ourselves fitter to enter on that path that leads to God and to free our spirit to gain and use those things which are far more precious than gold or precious stones. BASIL THE GREAT

Thus walking on this way with an awakened heart, not only do we escape the heavy cares of this world, but we are uplifted above ourselves and do taste of the divine sweetness. BL. ANGELA OF FOLIGNO

As long as you seek out and love the company of men of the world, Jesus Christ Whose delight it is to be with the children of men will not take any delight in you and will not give you any taste of the consolations with which He refreshes those who find all their joy in conversing with Him. Do not make friends with any persons except those whom you can help or those who can help you and animate you, by word and example, to love Jesus and live in His spirit. JEAN EUDES

Rejoice because you have reached the quiet and safe anchorage of a secret harbor. Many wish to come into this port, and many make great efforts to do so, yet do not achieve it. Indeed many, after reaching it, have been thrust out, since it was not granted to them from above.   BRUNO

A monastery is an academy of strict correction, where each one should allow himself to be treated, planed, and polished, so that, all the angles being effaced, he may be joined, united, and fastened to the will of God.
                                            FRANCIS DE SALES

Those of us who live in community must fight by the hour against all the passions and especially against these two: a mania for gluttony and bad temper. There is plenty of food for these passions in a community.
                                            JOHN CLIMACUS

Indeed I am sharper and more biting than pepper, and not infrequently when I preside over my Chapter I flare up over quite little matters. But they know that they have to endure the Bishop whom they have been given and so make a virtue of necessity and give way to me.   HUGH OF LINCOLN

For the superior is not to be obeyed because he is prudent, or kind, or divinely gifted in any other way, but for the sole reason that he holds the place of God and exercises the authority of Him Who says, "He who hears you hears me and he who despises you despises me."   IGNATIUS OF LOYOLA

Those who conscientiously fulfill the office of prior in these days ought to look not for rest, but must expect weariness and annoyance, a life of bitterness and misery . . . We fulfill our duty most perfectly by refusing to follow the timid suggestions of our sensitive natures.   BL. HENRY SUSO

When those who are ever ready to criticize do not usurp authority which they do not possess, as a rule they are very useful to the community because they cause everyone to be on the lookout.   ALPHONSUS LIGUORI

If ever there should be a monastery without a troublesome and bad-tempered member, it would be necessary to find one and pay him his weight in gold because of the great profit that results from this trial, when good use is made of it.   BERNARD OF CLAIRVAUX

Let the abbot aim at being loved rather than feared. He must not be worried nor anxious, neither should he be too exacting or obstinate, or jealous, or oversuspicious, for then he will never be at rest . . . Let him so temper all things so the strong may have their scope and the weak be not scared.   BENEDICT OF NURSIA

Be a John the Baptist to the incestuous, a Phineas to those who apostatize and go whoring, a Peter to liars, a Paul to blasphemers, a Christ to traders.

PETER MARTYR, *to a prioress*

Never will I suffer amongst us those souls without courage, those woman-ish hearts that can endure nothing. There must be nothing little amongst us. A religious must not be taken up with a headache, with those thousand aches and pains to which we are subject.   JULIE BILLIART

Each one should confidently make known his need to the other, so that he might find what he needs and minister to him. And each one should love and care for his brother in all those things in which God will give him grace, as a mother loves and cares for her son.   FRANCIS OF ASSISI

If the Blessed Virgin were on earth and wanted to become a nun, she would never be able to get into your convent, being a mere carpenter's wife, but the nuns of the Gesù would take her without any difficulty. This will show you in what favor you will be with the Queen of Heaven and her divine Son if you persist in such a spirit of wordly vanity.

ROBERT BELLARMINE, *to the nuns of San Giovanni*

What made me love the life of blessed Francis so much was the fact that it resembled the beginning and growth of the church. As the church began with simple fishermen and afterwards developed to include renowed and skilled doctors, so it was in the case of the order of Francis, showing that it was not founded by the prudence of men, but by Christ.

BONAVENTURE

MEN AND WOMEN

Because the Church exists in the world, and because of the unruly nature of sexual desire, traditional attitudes toward women have been full of contradiction and ambivalence. The saints themselves participate in this confusion. For the male saints, Woman has almost always figured as Temptation. This attitude, so closely connected with the effort to subdue the flesh for the sake of the spirit, is often held to have been responsible for repression of women within the Christian tradition.

There are a number of traditional elements in Christianity that have worked for woman's equality as well, however. Christ's own attitude was impeccable; he counted many women as his disciples and closest friends, chose a woman to proclaim the resurrection, and taught the sanctity of marriage. The veneration of Mary as the Queen of Heaven and earth, and the qualities of the great female saints certainly had a profound effect. Throughout the history of the Church, women have been able to enter Religious life, making them equal in the struggle for spiritual perfection, often in spite of powerful social forces.

I wish that men were as resolute as women.   BL. ANNE JAVOUHEY

I still can't understand why it's so easy for a woman to get excommunicated in Italy! All the time people seemed to be saying: "No, you mustn't go here, you mustn't go there; you'll be excommunicated." There's no respect for poor, wretched women anywhere. And yet you'll find the love of God much commoner among women than among men, and the women at the Crucifixion showed much more courage than the Apostles, exposing themselves to insult and wiping our Lord's face.   THÉRÈSE DE LISIEUX

If I had been a man I would have been a great preacher.   TERESA OF AVILA

The Indian woman has to work while the man quietly smokes opium. . . . How grateful we should be to Christianity, which has raised the dignity of woman, re-establishing her rights, unknown to pagan nations. Until Mary Immaculate, the Woman *par excellence*, foretold by the prophets, Dawn of the Sun of Justice, has appeared on earth, what was woman? But Mary appeared, the new Eve, true Mother of the Living, and a new era arose for woman. She was no longer a slave but equal to man; no longer a servant, but mistress within her own walls; no longer the object of disdain and contempt, but raised to the dignity of mother and educator, on whose knees generations are built up. All this we owe to Mary.

FRANCES XAVIER CABRINI

Through the wonderful providence of God's goodness, a woman's lips brought the news of life because in Paradise a woman's lips had dealt death.

GREGORY THE GREAT

Though I prefer learning coupled with virtue to all the treasures of kings, yet renown for learning when not united to a good life is nothing but glittering and notorious infamy, especially in a woman. Since erudition in a woman is a novelty and a reproach to masculine sloth, many will gladly attack it and impute to scholarship what is really the fault of nature, thinking from the vices of the learned to get their own ignorance esteemed as a virtue. . . . If a woman should add to eminent virture even moderate intellectual ability, I believe she will profit more surely than if she had acquired the riches of Croesus and the beauty of Helen.   THOMAS MORE

O, how wouldst thou have this wife of thine?—I would not have her gluttonous—and thou are ever at thy fegetelli: that is not well. I would have her active—and thou are a very sluggard. Peaceful—and thou wouldst storm at a straw if it crossed thy feet. Obedient—and thou obeyest neither father nor mother nor any man; thou deservest her not. I would not have a cock—well, thou are no hen. I would have her good and fair and wise and bred in all virtue—I answer, if thou wouldst have her thus it is fitting that thou shouldst be the same; even as thou seekest a virtuous, fair, good spouse, so think likewise how she would fain have a husband prudent, discreet, good, and fulfilled of all virtue.   BERNARDINO OF SIENA

Surely Almighty God should be greatly loved by women, inasmuch as He did not disdain to be born of a woman. A wonderful honor and dignity indeed did He hereby confer on them, for while it was not granted to man to be called the Father of God, yet this great distinction was bestowed upon a woman, that she should be the Mother of God.   HUGH OF LINCOLN

Arriving at Rome, he observed to his companions that he noticed that all the windows were closed, meaning by that that they would have to suffer many contradictions. He also said, "We must walk very carefully and hold no conversations with women, unless they are well known." What happened to Master Francis is very pertinent here. At Rome he heard a woman's confession and visited her occasionally to talk about her spiritual life. She was later found to be pregnant. But it pleased God that the responsible party was caught. The same thing happened to John Codure whose spiritual daughter was caught with a man.   IGNATIUS OF LOYOLA

Look at me. God's mercy has preserved me to this day in bodily virginity, but I confess that I have not escaped from the imperfection of being more excited by the conversation of young women than by being talked at by an old woman.   DOMINIC

It is a dangerous thing to associate with women, who but harm him who wishes to belong to God. One must avoid them and associate with them only when obedience demands.   JOSEPH OF COPERTINO

There are those who have avoided the one trap only to be caught in a vastly different one. For a careless slut with unkempt hair and slouching gait, clad in dirty rags, vulgar, ill-bred, coarse of speech, living a life of misery and scorn, friendless and forsaken by all, has begun by arousing a man's pity and ended by dragging him to utter perdition.   JOHN CHRYSOSTOM

This fair lovely word *Mother* it is so sweet and so kind itself that it may not verily be said of none but of Him, and to her that is the very Mother of Him and of all. To the property of Motherhood belongeth kind love, wisdom, and knowing; and it is good: for though it be so that our bodily forthbringing be but little, low, and simple in regard of our ghostly forthbringing, yet it is He that doeth it in the creatures by whom it is done.
        JULIAN OF NORWICH

St. Jerome and St. Augustine not only exhorted excellent matrons and most noble virgins to study, but also, in order to assist them, diligently explained the abstruse meanings of Holy Scripture and wrote for tender girls letters replete with so much erudition that nowadays old men who call themselves professors of sacred science can scarcely read them correctly, much less understand them.   THOMAS MORE

Some believe that woman will not rise again in their own sex but will all become men, for God made man out of the dust, but the woman out of the man. But those seem to me to hold the better opinion who have no doubt that both sexes will rise again.   AUGUSTINE OF HIPPO

VOCATIONS

A vocation is a call from God to a state of life in which one can reach holiness. More generally, all are called to communion with Christ, to peace, liberty, and sanctification. The idea of being called is not limited to the Religious life. Every role in life if carried out in the right spirit, say the saints, has an element of consecrated service.

Our Lord summoned Matthew by speaking to him in words. By an invisible, interior impulse which flooded his mind with the light of grace, the Lord instructed him to walk in his own footsteps. In this way, Matthew came to understand that Christ, who was summoning him away from earthly possessions, had incorruptible treasures of heaven in his gift.

BEDE THE VENERABLE

I opened the gospels at random, and the words my eyes fell on were these: "Then he went up on to the mountainside and called to him those whom it pleased to call; so these came to him." There it all was, the history of my life, of my whole vocation; above all of the special claims Jesus makes on my soul. He doesn't call the people who are worthy of it; no, just the people it pleases him to call.

THÉRÈSE DE LISIEUX

I read the heading of the letter which said, "The cry of the Irish," and at that very moment I heard the voice of those who were by the Wood of Voclut which is near the Western Sea, and this is what they cried, as with one voice, "Holy boy, we are asking you to come and walk with us again."

PATRICK OF IRELAND

And our Lord, seeking his workman among the multitude of those to whom he thus speaks, says again, "Who is the man that will have life and desireth to see good days?". . . What can be more agreeable, dearest brethren, than this voice of our Lord inviting us? Behold how in His loving kindness He shows us the way of life!

BENEDICT OF NURSIA

The thought occurred to him of joining the Carthusians of Seville. He could there conceal his identity so as to be held in less esteem, and live on a strictly vegetable diet. But, as the thought returned of a life of penance which he wanted to lead by going about the world, the desire of the Carthusian life grew cool, since he felt that there he would not be able to indulge the hatred he had conceived against himself.   IGNATIUS OF LOYOLA

Several friends had tried to persuade me to carry out the plan of overseeing the schools for the poor, but the idea never got hold of my mind and I had never the least thought of carrying it out. In fact, if I had ever suspected that the care I took of those first schoolmasters from pure charity would oblige me to remain with them for life I would have abandoned it instantly.   JEAN BAPTISTE DE LA SALLE

Perhaps you and I will find ourselves soldiers of the same regiment, travelers on the same road, bound for the same destination. May His holy will, not ours, be done! Leave your future in His hands, in the heart of Jesus made man. Remember that he too was once a young man, for Jesus Christ is the God-child, the God-youth, the God-man, the God of all ages.

                                                  BL. THEOPHANE VENARD

He need have no fear of error in believing that God is calling him to contemplation, regardless of what sort of person he is now or has been in the past. It is not what you are nor what you have been that God sees with his all-merciful eyes, but what you desire to be.

                                ANONYMOUS, *The Cloud of Unknowing*

I do implore you, dearest Father, to take the greatest care never to influence my little brother, either by word or in any way whatsoever, to enter upon the ecclesiastical state. Were he to decide upon it without a very special vocation, above all were he to take it up as a career from purely human motives of self interest, he would be guilty of a dreadful sacrilege, and for you as well as for himself the step would prove the most deplorable disaster.   BL. JOHN GABRIEL PERBOYRE

It is not so easy as you may think to take the habit against your parents' wishes, for though you may be resolved to do so now, I doubt whether you are so holy as not to feel unhappy afterwards at being in disgrace with your father. The best way is to pray about the matter to our Lord and leave it to Him. He can change hearts and find means to bring it about.

                                                  TERESA OF AVILA

My young clerics were full of good qualities—although they were so undisciplined, they were good workers, good hearted, and morally sound. I always thought that when the first fires of youth were quenched they would prove invaluable, and I was right. But if I had insisted on perfection I would have achieved nothing.   JOHN BOSCO

It is most laudable in a married woman to be devout, but she must never forget that she is a housewife; and sometimes she must leave God at the altar to find Him in her housekeeping.   FRANCES OF ROME

Each state of life has its special duties; by their accomplishments one may find happiness in the world as well as in solitude, for not all are called to separate themselves from the society of men, like John the Forerunner, and to lead in the desert a solitary life.   NICHOLAS OF FLÜE

Our Lord has created persons for all states in life, and in all of them we see people who have achieved sanctity by fulfilling their obligations well.

ANTHONY MARY CLARET

Just as every sort of gem when cast in honey becomes brighter and more sparkling, so each person becomes more acceptable and fitting in his own vocation when he sets that vocation in the context of devotion. Through devotion family cares become more peaceful, mutual love between husband and wife becomes more sincere, the service we owe the prince becomes more faithful, and our work, no matter what it is, becomes more pleasant and agreeable.   FRANCIS DE SALES

My longing to save souls grew from day to day; it was as if I heard our Lord saying to me what he had said to the Samaritan: "Give me some water to drink."   THÉRÈSE DE LISIEUX

WORKS AND WORK

The saints perform the scriptural works of mercy, corporal and spiritual, which include feeding the hungry, giving drink to the thirsty, clothing the naked, visiting the imprisoned, sheltering the homeless, visiting the sick, and burying the dead. The traditional works of spiritual mercy are counseling the doubtful, instructing the ignorant, admonishing sinners, comforting the afflicted, forgiving offenses, bearing wrongs patiently, and praying for the living and the dead.

"Faith without works is dead" (James 2:20), but faith and sanctifying grace are also necessary for salvation.

We have a long way to go to Heaven, and as many good deeds as we do, as many prayers as we make, and as many good thoughts as we think in truth and hope and charity, so many paces do we go heavenward.

BL. RICHARD ROLLE

I believe my days are few. I feel such an extraordinary desire to work and to serve God, I feel it so passionately, that I cannot believe God would have given it to me if He did not mean to take me away at once.

ALOYSIUS GONZAGA

We must also be on our guard against exalting Divine grace so much as to make our hearers no longer believe themselves free: we must speak of it as the glory of God requires, that we may not raise doubts as to liberty and the efficacy of good works.   IGNATIUS OF LOYOLA

Do not suppose that God has any need of our works; what he needs is the resoluteness of our will.   TERESA OF AVILA

God does place more value on good will in all we do than on the works themselves. Whether we give ourselves to God in the work of contemplation, or whether we serve the needs of our neighbor by charitable works, we accomplish these things because the love of Christ urges us on.

LAWRENCE GIUSTINIANI

The Servant of Charity must go to bed each night so tired from work that he will think he has been beaten. BL. LOUIS GUANELLA

The true apostolic life consists in giving oneself no rest or repose.

CAMILLUS DE LELLIS

We will lie down for such a long time after death that it is worthwhile to keep standing while we are alive. Let us work now; one day we shall rest.

BL. AGOSTINA PIETRANTONI

The chief thing is to take the burden on one's shoulders. As you press forward, it soon shakes down and the load is evenly distributed.

JOHN BOSCO

I don't know how I could have managed to write so many different books. You must have done it, Lord. I know that this is putting it badly; I haven't written anything, you have done it all. My God, you made use of me, a worthless instrument without the knowledge, talent, or time to do all this. But, unknown to me, you were giving me all the help I needed.

ANTHONY MARY CLARET

What labor my writing cost me, what difficulties I went through, how often I despaired and left off, and how I began to learn again, both I who felt the burden can witness and they also who lived with me. JEROME

Your fellow men are marvelously enriched by your contemplation, even if you may not fully understand how; the souls in purgatory are touched, for their suffering is eased by the effects of this work; and, of course, your own spirit is purified and strengthened by this contemplative work more than by all others put together. ANONYMOUS, *The Cloud of Unknowing*

Until the soul reaches the state of union of love, she should practice love in both the active and contemplative life. Yet once she does arrive, she should not become involved in other works and exterior exercises that might be of the slightest hindrance to the attentiveness of love toward God, even though the work might be of great service to God. For a little of this pure love is more precious to God and the soul and more beneficial to the church, even though it seems one is doing nothing, than all these other works put together. JOHN OF THE CROSS

Hell is full of the talented, but Heaven of the energetic.

JEANNE FRANÇOISE DE CHANTAL

A good work talked about is a good work spoilt. VINCENT DE PAUL

Good works are the most perfect when they are wrought in the most pure and sincere love of God, and with the least regard to our own present and future interests, or to joy and sweetness, consolation or praise.

JOHN OF THE CROSS

PRIESTS

A sacrament is a rite which acts as a visible sign and channel for supernatural grace. The sacrament of Holy Orders confers the power of consecrating and offering the body and blood of Christ and of remitting or retaining sins. The priest's role is to mediate between the people and God. Ordinary priests can perform six of the seven sacraments; Baptism, Confirmation, Holy Eucharist, Penance, Annointing of the Sick, and Matrimony. Holy Orders can be conferred only by a bishop.

Clergy, like the Church, certainly have a mundane as well as a divine aspect, but the saints, as a rule, dwell on the supernatural aspects of the office or else on opportunities afforded by clerical shortcomings to practice the virtues of obedience, charity and humility.

The most high and infinitely good God has not granted to angels the power with which he has invested priests.   JOHN CHRYSOSTOM

Oh venerable dignity of priests, in whose hands the Son of God becomes incarnate anew and He formerly became incarnate in the womb of Mary.
                                                                        AUGUSTINE OF HIPPO

Oh! how great is the priest. The priest will only be understood in heaven. Were he understood on earth, people would die, not of fear, but of love.
                                                                        JOHN VIANNEY

The Lord gave me and still gives me such faith in priests who live according to the manner of the Holy Roman Church because of their order, that if they were to persecute me, I would still have recourse to them. And if I possessed as much wisdom as Solomon had and I came upon pitiful priests of this world, I would not preach contrary to their will in the parishes in which they live.   FRANCIS OF ASSISI

Like a farmer tending a sound tree, untouched by ax or fire because of its fruit, I want to serve you good people, not only in the body, but also to give my life for your well-being.   EUSEBIUS OF VERCELLI

Priest at the altar, priest in the confessional, priest among my boys, and priest in the King's palace or his ministers' offices: I will be nothing but a priest.   JOHN BOSCO

I offered my neck, then, to the yoke of Christ, and now I see myself engaged on tasks that are too great for my deserts and for my understanding. I know that I have now been admitted and accepted into the secret shrines of the highest God, that I partake of heavenly life, and that, brought nearer to God, I dwell in the very spirit, body, and illumination of Christ. With my puny mind I can as yet scarcely understand the sacred burden I bear, and, aware of my weakness, I tremble at the weight of my task.

PAULINUS OF NOLA

St. Peter and St. John both say that all good Christians are like priests. Now it is the function of priests to offer sacrifice. Thus, as Jesus Christ the High Priest offered Himself as a victim for the glory of the Eternal Father and in satisfaction for our sins, so we, too, as true Christians, are priests and as such should offer ourselves for the glory of God and in satisfaction for our sins and those of all the nation.   ANTHONY MARY CLARET

Do you not shudder to admit a soul like mine to so sacred a ministry, to raise to the priestly dignity one clad in foul garments, a man whom Christ has excluded from the banquet of guests? A priest's soul should be resplendent with light, like a torch illuminating the whole world, while mine is so obscured by the dark mist rising from an impure conscience that, being always dispirited, it never dares to look its divine Master confidently in the face.   JOHN CHRYSOSTOM

Do you uproot in the garden of Holy Church the malodorous flowers, full of impurity and avarice, swollen with pride: that is, the bad priests and rulers who poison and rot that garden. Ah me, you our Governor, do you use your power to pluck out those flowers! Throw them away, that they may have no rule! Insist that they study to rule themselves in holy and good life.   CATHERINE OF SIENA, to Pope Gregory XI

I take this opportunity to tell you that when I consider the many distractions and much worldly pomp of my present position, I envy bishops, because I think their state is safer and more like that of Religious, and I find, too, the Calendar full of sainted bishops but can discover only one cardinal, St. Bonaventure, and he lived as a cardinal only a few years.

ROBERT BELLARMINE

Listen my brothers: if the blessed Virgin is so honored, as it is right, since she carried him in her most holy womb; if the blessed Baptist trembled and did not dare to touch the holy head of God; if the tomb in which he lay for some time is so venerated, how holy, just, and worthy must be the person who touches Him with his hands, receives Him in his heart and mouth, and offers Him to others to be received.   FRANCIS OF ASSISI

If her priests are saints, what good they are able to do! But whatever they are, *never speak against them.*   JOHN VIANNEY

TEACHING
AND PREACHING

Teaching and preaching count as spiritual works of mercy since their practitioners instruct the ignorant, admonish sinners, counsel the doubtful, and comfort the sorrowful. Christ himself comissioned the first preachers, telling them to "go into all the world and preach the gospel to every creature" (Mark 16:15) and to "teach all nations" (Matthew 28:19).

The saints say that God is willing to be the teacher of all who are able to listen, through his gifts of discernment, inspiration, illumination, and guidance. Still, human instruments are useful, and they should be prepared to be flexible and able to adapt to the needs of the students.

The stunning realism which seems to be a consistent feature of sainthood is often evoked by the teaching role as when we see the prodigiously learned St. Jerome willing to accommodate the educational needs of a little girl, and St. John Bosco urging priests to climb down from the pulpit and play with the children if they want to inspire learning that is based on love and trust. Kindness and the power of example are constantly stressed.

The saints in general do agree that simplicity and inspiration are the "secrets" of effective preaching. Integrity and holiness of life are also helpful. Grace is necessary for the hearer as well as the preacher.

Our Lord doesn't need to make use of books or teachers in the instruction of souls; isn't he himself the Teacher of all Teachers, conveying knowledge with never a word spoken? For myself, I never heard the sound of his voice, but I know that he dwells within me all the time, guiding me and inspiring me whenever I do or say anything. A light, of which I'd which I'd caught no glimmer before, comes to me at the very moment when it's needed.   THÉRÈSE DE LISIEUX

At this time God treated him just as a schoolmaster treats a little boy when he teaches him. This perhaps was because of his rough and uncultivated understanding, or because he had no one to teach him, or because of the firm will God himself had given him in His service. But he clearly saw, and always had seen, that God dealt with him like this.

IGNATIUS OF LOYOLA

Seeing these innocent souls at close quarters, I realized what a mistake it is not to train them from the very start when they are like wax to receive impressions for better or for worse. There are so many souls which would attain sanctity if only they were well directed. THÉRÈSE DE LISIEUX

When I pointed out to him in my room what unheard-of nonsense he had been talking, he edified me by his humility and obedience. So I told him that, on the following day, he must go into the pulpit and declare that the statements he had made in his sermon were slips of the tongue due to rhetorical exaggeration. This he did most thoroughly, and I took the opportunity to give him a good brotherly reproof, putting him in mind of the rule of St. Francis about simplicity in teaching. Then, to sweeten the medicine, I sent him some trout. ROBERT BELLARMINE

The abbot in his teaching should always observe that apostolic rule which saith, "Reprove, entreat, rebuke." That is to say, as occasions require he ought to mingle encouragement with reproofs. Let him manifest the sternness of a master and the loving affection of a father. BENEDICT OF NURSIA

If you will send your tiny Paula, I promise myself to be both teacher and foster-father to her. Old as I am, I shall carry her pickaback on my shoulders and train her stammering speech. Far more glorious than that of the worldly philosopher shall my task be, for I shall instruct, not a Macedonian king to die one day of Babylonian poision, but a handmaid and bride of Christ to be prepared for the kingdom of heaven. JEROME

You can do nothing with children unless you win their confidence and love by bringing them into touch with oneself, by breaking through all the hindrances that keep them at a distance. We must accommodate ourselves to their tastes, we must make ourselves like them. JOHN BOSCO

I imagine that the angels themselves, if they came down as schoolmasters, would find it hard to control their anger. Only with the help of the Blessed Virgin do I keep from murdering some of them. BENILDUS

What is nobler than to rule minds or to mould the character of the young? I consider that he who knows how to form the youthful mind is truly greater than all painters, sculptors, and all others of that sort. JOHN CHRYSOSTOM

It is chiefly by asking questions and in provoking explanations that the master must open the mind of the pupil, make him work, and use his thinking powers, form his judgement, and make him find out for himself the answer.   JEAN BAPTISTE DE LA SALLE

All who undertake to teach must be endowed with deep love, the greatest patience, and, most of all, profound humility. They must perform their work with earnest zeal. Then through their humble prayers, the Lord will find them worthy to become fellow workers with him in the cause of truth.
                                                        JOSEPH OF CALASANZA

You are a candle worthily set on the candlestick of the Church; from your seven-branched lamp you pour out light, nourished by the oil of joy, far and wide over Catholic cities. You dispel the mist of heretics however thick it lies, and with the brilliance of your brightening words uncover the light of truth from the disorder of darkness.   PAULINUS OF NOLA

When St. Paul's voice was raised to preach the gospel to all nations, it was like a great clap of thunder in the sky. His preaching was a blazing fire carrying all before it. It was the sun rising in full glory. Infidelity was consumed by it, false beliefs fled away, and the truth appeared like a great candle lighting the whole world with its brilliant flame.
                                                        BERNARDINO OF SIENA

That the preaching of these men was indeed divine is clearly brought home to us when we consider how else could twelve uneducated men, who lived on lakes and rivers and wastelands, get the idea for such an immense enterprise? How could men who perhaps had never been in a city or a public forum think of setting out to do battle with the whole world?
                                                        JOHN CHRYSOSTOM

Do you really think I said fine things? An hour before preaching I was not prepared, but that hour I spent before the tabernacle. Then I said to Our Lord, "Let us go preach." It was Our Lord who spoke to you.
                                                        PETER JULIAN EYMARD

Happy the man whose words issue from the Holy Spirit and not from himself.   ANTHONY OF PADUA

A preacher is like a trumpet which produces no tone unless one blows into it. Before preaching, pray to God, "You are the spirit and I am only the trumpet, and without your breath I can give no sound."
                                                        JOSEPH OF COPERTINO

When he stood in their midst to present his edifying words, he went completely blank and was unable to say anything at all. This he admitted with true humility and directed himself to invoke the grace of the Holy Spirit. Suddenly he began to overflow with eloquence.

BONAVENTURE, *Life of St. Francis*

When I see the need there is for divine teaching and how hungry people are to hear it, I am atremble to be off and running throughout the world, preaching the word of God. I have no rest, my soul finds no other relief than to rush about and preach. ANTHONY MARY CLARET

I must preach so that the most illiterate laborer can understand me.

ALPHONSUS LIGUORI

Jesus Christ, in his infinite Wisdom, used the words and idioms that were in use among those whom he addressed. You should do likewise.

JOSEPH CAFASSO

So long as we are still in this place of pilgrimage, so long as men's heart's are crooked and prone to sin, lazy and feeble in virtue, we need to be encouraged and roused, so that brother may be helped by brother and the eagerness of heavenly love rekindle the flame in our spirit which our everyday carelessness and tepidity tend to extinguish. BL. JORDAN OF SAXONY

If teaching and preaching is your job, then study diligently and apply yourself to whatever is necessary for doing the job well. Be sure that you first preach by the way you live. If you do not, people will notice that you say one thing, but live otherwise, and your words will bring only cynical laughter and a derisive shake of the head. CHARLES BORROMEO

Let the meaning of your words shine forth, let understanding blaze out from them. Let no word escape your lips in vain or be uttered without depth of meaning. AMBROSE

Teaching unsupported by grace may enter our ears, but it never reaches the heart. When God's grace does touch our innermost minds to bring understanding, then his word which is received by the ear can sink deep into the heart. ISIDORE OF SEVILLE

Take particular care never in your sermons to denounce any man who holds a public charge, for there is danger that such people may become worse instead of better if reprehended from the pulpit. . . . If you rebuke important or rich persons in good round terms, I am afraid they only lose patience and turn hostile. FRANCIS XAVIER

COUNSELING

The saints' terminology differs from that of modern psychology, but it is easy to see that in their struggles to enlighten and deliver souls, or simply to give good advice in a way that might ensure that it will be helpful, they are dealing with the same human dilemmas, pain, conflict, and danger. In counseling, something must be present in the attitude of the confessor, director, helper, therapist, or counselor that adds a healing element to the relationship or there can be no change for the better. The gifts of discernment, wisdom, and understanding are particularly useful, as is the supreme virtue of charity.

The better friends you are, the straighter you can talk, but while you are only on nodding terms, be slow to scold.　　　　　　　　　FRANCIS XAVIER

In dealing with another, we should take a cue from the enemy who wishes to draw a man to evil. He goes in via the way of the man he wishes to tempt, but comes out his own way. We may thus adapt ourselves to the inclinations of the one with whom we are conversing, adapting ourselves in our Lord to everything, only to come out later with the good accomplished to which we had laid our hand.　　　　　　　　　IGNATIUS OF LOYOLA

Men's bodies, as you know, are of very various temperaments, and there is just as great a dissimilarity in the constitution of their minds, for God bestows very different gifts upon different individuals. He does not lead all by the same path, therefore it is impossible to specify any particular devotion as the most suitable. Some have no special attraction to any one form of devotion, and they ought to consult someone as to their interior dispositions, so as to know whether they should allow themselves to be led by motives of love or fear, of sadness or joy; and how to apply the remedies most suitable to their needs.　　　　　　　　　JOHN OF AVILA

We should analyze the nature of our passions and of our obedience and choose our director accordingly. If lust is your problem, do not pick for your trainer a worker of miracles who has a welcome and a meal for everyone. Choose instead an ascetic who will reject any of the consolation of food. If you are arrogant, let him be tough and unyielding, not gentle and accommodating. Do not search for someone gifted with foreknowledge and foresight, but rather for one who is truly humble and whose character and dwelling complement our weaknesses.   JOHN CLIMACUS

When God has designs on a soul, He gives the confessor the necessary light to know that soul and to guide her.   PETER JULIAN EYMARD

The Lord our God has granted me the grace of knowing the heart of people as if I were reading a book.   ANTHONY MARY CLARET

Take care not to frighten away by stern rigor poor sinners who are trying to lay bare the shocking state of their souls. Speak to them rather of the great mercy of God, and make easy for them what is at best a difficult task. Be especially gentle with those who from weakness of age or sex have not the courage to confess the ugly things they have done. Tell them that whatever they have to say will be no news to you. Sometimes people are helped by your telling them about your own lamentable past.   FRANCIS XAVIER

Remember that the confessor is a father who is eager to do all he can for you and protect you from all possible harm. Never fear that you will lose his respect when confessing serious sins, or that he will reveal them to others. No matter what may happen to him, the confessor may never avail himself of any information received in the confessional. Should he stand to lost his life, he can not and will not tell anyone at all even the slightest thing hear in the confessional. Moreover, I can assure you that the more sincere and trusting you are with him, the more his confidence in you will grow and the better he will be able to give you that advice and counsel which he deems most necessary and useful for your soul.   JOHN BOSCO

They should rest content with one good confessor and trust themselves entirely to his judgement and direction. At the Day of Judgement, while the confessor must give an account of his direction of the penitent, the penitent himself is shielded behind his obedience and submission to his confessor.   JOHN VIANNEY

Father Cafasso had spent a whole week instructing and encouraging forty-five notorious criminals in a large cell, in preparation for a feast of the Virgin Mary. Almost all of them had promised to go to confession on the eve of the feast day. But when the day arrived, no one had the courage to be the first one, whether because of human respect, the guile of the devil, or for some other reason.

Father Cafasso renewed his request, briefly recalling the instructions of the preceding days and reminding them of their promise, but in vain. What was he to do?

Supernatural love is ingenious, and Father Cafasso found a solution. Laughingly he approached one of them, the tallest and strongest among them. Without uttering a word, he grabbed the man's long, thick beard in his frail hands. At first the prisoner thought that Father Cafasso was jesting, so he only said, with as much courtesy as one might expect from one of his kind, "Take all of me, but leave me my beard."

"I'm not letting go until you come to confession."

"But I'm not coming."

"Then I won't let go of you."

"I don't want to go to confession."

"Talk all you want, you're not going to get away from me. I won't let go of you until you've made your confession."

"I'm not prepared."

"I'll help you."

The convict could very easily have shaken off Father Cafasso's hold, but either because of respect for him, or, better yet, because of God's grace, it is a fact that the convict yielded and allowed Father Cafasso to lead him to a corner of the large cell. The priest sat down on a straw mattress and prepared him for confession. Suprisingly, in a few moments, the convict was deeply moved, and amid tears and sighs, was just barely able to finish the recitation of his sins.

Then something wonderful took place. This man, who previously had refused, amid blasphemies, to go to confession, began to tell his fellow inmates that never in all his life had he been as happy as right then. He was so enthused that he talked them all into making their confession.

Whether one wishes to interpret this incident, one from among many, as a miracle of God's grace, or as a miracle of Father Cafasso's charity, he cannot but see in it the hand of God.   JOHN BOSCO, *of St. Joseph Cafasso*

PAGANISM

Saints have had to confront both idolatrous pre-Christian and secularist post-Christian forms of paganism. Selections dealing with earlier forms of paganism seem filled with vitality and promise. By contrast, the secularism of our own era appears to present a grimmer challenge.

In days of old, evil spirits appeared in various guises and defiled women and corrupted boys, and made a show of horrors. Men were seized with dread and failed to understand that they were wicked spirits; instead they called them gods and addressed them by all the titles that the demon bestowed on himself.
JUSTIN MARTYR

If hereafter you wish to come to the Blessed City, avoid the company of demons. They can be satisfied with the worship of rogues, to whom immorality is pleasing. Therefore let your gods be erased from your worship by a Christian cleansing, just as the stage-players have been removed from honorable society by the action of the censors.
AUGUSTINE OF HIPPO

They who worship images are slaves to demons. Do you address as lords and greet with bowed head men whom you see enslaved to wood and stone? They worship gold and silver under the name of gods, and their religion is one that corrupt greed also loves.
PAULINUS OF NOLA

Reject absolutely all divination, fortune-telling, sacrifices to the dead, prophesies in groves or by fountains, amulets, incantations, sorcery (that is wicked enchantments), and all those sacrilegious practices which used to go on in your country.
GREGORY III, *to the Germans*

These same men, who shave their hair and eyebrows if they are initiated into the mysteries of Isis, cry out against the indignity if anyone changes his garment for the sake of the holy religion of Christ. So much consideration for falsehood and contempt for the truth!
AMBROSE

When an Indian dies, all the friends are called to weep over the corpse, whether they want to or not. They even have to chant their grief in a more or less monotonous strain like this: "You were very good, oh, oh, oh! You had a lovely house, ah, ah, ah! . . ." They hold the strange belief that when the corpse is on earth it needs nothing, but if it goes to hell, it has to be provided with bread and water. Naturally, the dead man does not return to take bread and water; consequently the tribe concludes he has gone to heaven and makes merry over it, partaking of a great banquet.

FRANCES XAVIER CABRINI

Among you the idols are being abolished, but our faith is spreading everywhere. . . . By your beautiful language you do not impede the teaching of Christ, but we, calling on the name of Christ crucified, chase away all the demons you fear as gods. Where the sign of the cross occurs, magic loses its power, and sorcery has no effect.   ANTHONY OF EGYPT, *to the Greeks*

In your intercourse with men, you will encounter much prejudice, many strange ideas and perversions of the truth, for society in Europe has become thoroughly corrupt. I do not mean to say that there were not plenty of bad people in olden times just as there are now, for man is ever the same. But formerly there were certain social canons and conventions that none but the really profligate disregarded. Religion was the accepted foundation of society, and God gives life to nations as well as to individuals. Now all these safeguards are removed or ignored.   BL. THEOPHANE VENARD

HERESY

To become a heretic one must first be a baptized, professing Christian who refuses to accept a truth revealed by God and proposed for belief by the Church. Those who cease to profess Christianity are apostates rather than heretics.

The history of the Church is filled with a bewildering abundance of heresies, beginning with the Judaizers, the Gnostics, and a variety of anti-Trinitarians. Modern heresies tend to take the form of denial of the supernatural as in Naturalism or Humanism. Americanism and Modernism suggest that truth be adjusted to fit the temper of the times.

He said that we must be slain and hated for His name's sake, and that many false prophets and false Christs would come forward in His name and would lead many astray. And this is the case. For many have taught what is godless and blasphemous and wicked, falsely stamping their teaching with His name, and have taught what has been put into their mind by the unclean spirit of the devil, and teach it until now. **JUSTIN MARTYR**

Avoid heretics like wild beasts; for they are mad dogs, biting secretly. **IGNATIUS OF ANTIOCH**

Do not defile yourselves with the Arians, for that teaching is not from the apostles, but from the demons, and from their father, the devil; Indeed, it is infertile, irrational, and incorrect in understanding, like the senselessness of mules. **ANTHONY OF EGYPT**

Sensual living men of the world who feel that the Church's requirements for the proper amendment of their lives are too burdensome quickly and easily follow after these heretics and fiercely support them. All because they imagine that these heretics will lead them by a smoother path than Holy Church. **ANONYMOUS**, *The Cloud of Unknowing*

Concerning your statement that you have heretics with whom you constantly disagree and whom you wish to recall to the Catholic faith, I praise your zeal but reprehend your audacity, for, as the divine scripture says, "Whoever has touched pitch will be defiled by it." Thus beware, beloved son, lest . . . you succumb, conquered ignominiously by heretical persuasions; for while you labor to rescue them from the error of death, they themselves stay awake to submerge you in the abyss of error.

ISIDORE OF SEVILLE

In Him, therefore, do I understand and possess all truth that is in heaven and earth and hell, and in all creatures; and so great is the truth and the certainty that were the whole world to declare the contrary I would not believe it, yea, I should mock at it.   BL. ANGELA OF FOLIGNO

Hear God's word, ye fishes of the sea and river, since the stiff-necked Cathars will not hearken to it!   ANTHONY OF PADUA

In condemning us, you condemn all your own ancestors—all the ancient priests, bishops, and kings—all that was once the glory of England, the isle of Saints, and the most devoted child of the See of Peter. For what have we taught, however you may qualify it with the odious name of treason, that they did not uniformly teach?   EDMUND CAMPION

Christ really suffered as he really raised himself. Some unbelievers say that he suffered in appearance only. Not so—they themselves are mere apparitions. Their fate will be like their opinions, for they are unsubstantial and phantom-like.   IGNATIUS OF ANTIOCH

MISSIONS

Mission, or "sending," is the essence of Christianity, beginning with the Trinity itself. The Father sends the Son on a mission to the world, and the Son sends the Holy Spirit. Christ's last command to his Apostles is "Go ye therefore and teach all nations" (Matthew 28:19). Mission is thus the purpose of vocation.

In all ages the missionary is as likely as not to be resented for undermining the old values, as Jesus was, or as a foreign intruder, or in current terminology, as an agent of "cultural imperialism." The risk, and indeed the probability, of martyrdom has always been recognized and eagerly sought by many of the saints as the most literal imitation of Christ and the fullest expression of heroic devotion. Saint Francis and Saint Anthony of Padua were keenly disappointed, for example, when they escaped death at the hands of the Saracens.

Risks are counted as nothing in exchange for the chance of experiencing the joy of winning souls.

Hearing, to our great distress, that certain peoples in Germany on the eastern side of the Rhine are wandering in the shadow of death, at the instigation of the ancient Enemy, and are still in slavery to the worship of idols, we have determined to send our brother Boniface into that country to bring them eternal life.   GREGORY II

I want to be a warrior, a priest, a doctor of the church, a martyr. I want to go to the ends of the earth to preach Your name, to plant Your glorious Cross on pagan shores.   THÉRÈSE DE LISIEUX

Pour into their untaught minds the preaching of both the Old and the New testament in the spirit of virtue and love and sobriety and with reasoning suited to their understanding.   BONIFACE

An apostolic missionary must have both heart and tongue ablaze with charity.   ANTHONY MARY CLARET

Who, dear brother, can possible describe the great joy of believers when they have learned what the grace of the Almighty God and your own co-operation have done for the Angles? They abandoned the errors of darkness and were suffused with the light of holy faith. With full awareness, they trampled on the idols which they had previously adored with savage fear. They are now committed to Almighty God.   GREGORY I THE GREAT

You beg for advice as to whether it is permitted to flee from the persecution of the heathen or not. We give you this wholesome counsel: so long as it can be done, and you can find a suitable place, carry on your preaching, but if you cannot endure their assaults you have the Lord's authority to go into another city.   ZACHARIAS

What consolation it would give the Church to see in her fold a whole people as notable and attractive as the Chinese.   BL. JOHN GABRIEL PERBOYRE

The people we have met so far is the best that has until now discovered; and it seems to me that among heathen peoples no other will be found to surpass the Japanese.   FRANCIS XAVIER

The best of news comes from Molucca: John Beira and his companions are constantly in peril and in danger of death, and the result is great progress in the Christian religion.   FRANCIS XAVIER

The field of the Lord, which had been lying fallow, bristling with the thorns of unbelief, has received the plowshare of your instruction and is bringing forth an abundant harvest of true belief.   GREGORY II

Voracious wolves have swallowed up the flock of the Lord which was in-creasing nicely as a result of hard work, and the sons of the Irish and the daughters of lesser kings were monks and virgins of Christ—I cannot count them!   PATRICK OF IRELAND

The bishop gave the Indian chief a crucifix which he accepted with respect. Later when he entered one of the stores in St. Louis, the shopkeeper, anx-ious to see if the crucifix was treasured, offered to give in exchange a fine saddle, then liquor, and finally a large sum of money. Each time the chief refused, saying that never would he give away what he had received from the "one who speaks to the Author of Life."   BL. PHILIPPINE DUCHESNE

I arrived back on Sunday evening, having heard very bad news about the Cape Comorin Christians. The Badagas, I learned, were carrying them off as slaves, and the Christians, to save themselves, had taken refuge on some rocks out in the sea, where they are now dying of hunger and thirst. The winds were so contrary that when I set out with twenty *tonis* to the succor of the fugitives, that neither by rowing nor by towing the boats with ropes from the shore could we reach the Cape. When the winds subside we shall have another try to help the poor souls whom it is the most pitiful thing in the world to see.   FRANCIS XAVIER

Our lives hang upon a single thread. Apart from the fact that your cabin is merely, so to speak, a thing of straw, and may be burned down at any moment, the ill will of the natives is enough to keep us in a state of almost perpetual fear. A malcontent may set you on fire, or choose some lonely spot to split your skull open. And then you are held responsible for the barrenness of the earth on pain of your life: you are the cause of droughts; if you cannot make rain, they go so far as threatening to do away with you. I leave it to you to imagine if we have any grounds for feeling secure.

                                                  JEAN DE BREBEUF

The convent which sheltered us before has been destroyed by the pagans, who got wind of our being there. We had just time to escape between two double walls about a foot wide. We saw through the chinks the band of persecutors, with the mayor at their head, garroting five or six of the oldest nuns who had been left behind.   BL. THEOPHANE VENARD

At the third village, we met four other Hurons newly-captured and muti- lated like ourselves. I managed to instruct and baptize them—two with the dew which I found quite abundant in the tall leaves of Turkish corn whose stalks they gave us to chew, the other two at a brook which we passed on our way to another village.   ISAAC JOGUES

You might ask why don't we go mad? Always shut up in the thickness of two walls, with a roof one can touch with one's hand, our companions spiders, rats, and toads, always obliged to speak in a low voice, "like the wind," as the Annamites say, receiving every day the most terrible news of the torture and death of our fellow missionaries, and worse still, of their occasional apostasy under torture. It requires, I own, a special grace not to be utterly discouraged and cast down.   BL. THEOPHANE VENARD

"Well, and is that all?" someone will exclaim. "Do you think such argu- ments are likely to quench the fire that consumes me? All you have told me seems to me nothing in comparison with what I am ready to endure for God. If I knew a place on earth where there was more to be suffered, I would go there." Ah. you, to whom God has given this desire and this light, come. come, dear Brother. It is workmen like you we want out here.

                                                  JEAN DE BREBEUF

As regards certain customs of these savages, do not try to belittle them, but rather, after the example of the Church in ancient times amidst pagan peoples, try to sanctify such customs, provided they are not harmful to soul or body.

BL. MICHAEL RUA

TRAVELS

The saints traveling—observing local customs, suffering upset stomachs in foreign inns, refusing to be intimidated by awful reports of the way foreigners will be treated in France, shopping, and being thrilled and surprised at courteous treatment—are the saints at their most human. They look on strange sights with divine aplomb, and are looked at in their turn, as at Mother Cabrini's Eskimo banquet, or in John Gabriel Perboyre's encounter with the young Chinese man who admired his nose.

Yielding to a secret inspiration of the soul I visited this country, England, that was once the island of saints and lost its faith, alas, through the pride and passion of its king. Pray, dear daughters, pray much for the conversion of England. It breaks one's heart to see this country deprived of the true faith. England has all the qualities that make it worthy to be a portion of Christ's fold. Her only fault is that of having but half the faith.

FRANCES XAVIER CABRINI

This evening I was watching a beautiful sunset on the English coast while the moon rose on the French side of the Channel. I couldn't help thinking about England, where the Sun of Truth has so long been darkened. I pray for this country with all my heart. England could do so much for the good cause if she would only make it her own. Britannia rules the waves, but she sows error in the lands under her flag. Let us pray that all this may be changed.

BL. THEOPHANE VENARD

It must be a rare sight for English people to see a priest in his cassock, for when we went ashore, men, women and children looked at us in amazement. Some of the little ones were fairly frightened and ran away. One of the men was curious enough to touch the cassock and examine the buttons. Then they burst out laughing, and that so naively that we laughed too.

Evidently they are very much like the Chinese in some ways, curious to the point of incivility and with very little sense in their mockery.

BL. THEOPHANE VENARD

One young pagan was admitted into a room where I was seated. He took up his position in front of me and studied me with as close an attention as if he intended to paint my portrait; he then retired perfectly satisfied, so he said, at having seen a European nose. It was his heart's desire to behold this marvel at least once during his lifetime.   BL. JOHN GABRIEL PERBOYRE

The Eskimo's manner of taking food is very strange. If you are invited by some great personage, such as the head of the tribe, you must not imagine you are going to eat a piece of salmon or roast codfish, in which these coasts abound. In front of the head of the family you see two plates, one with the dressed meats and the other empty. Now his work begins, and this must be very hard, for he chews all the food which is given to the guests. When this is done, it is placed on plates and handed around accordingly. This ceremony over, all the guests eat of this well-prepared dish.

FRANCES XAVIER CABRINI

When he arrived at Barcelona, all who knew him tried to dissuade him from passing over to France because of hostilities. They recounted many instances of atrocities, even going so far as to say the French roasted Spaniards on spits. But he saw no reason for being afraid.   IGNATIUS OF LOYOLA

We traveled for twenty-eight days through uninhabited country, and their food ran out and famine prevailed over them and one day the captain turned to me and said, "What now, Christian? You say that your God is great and almighty; why then can you not pray for us?" . . . The next thing was a herd of pigs appeared on our route in front of our eyes.

PATRICK OF IRELAND

I have always known that the Lord was my fuel; but on this trip all the rest knew it too. They could see that I hardly ate or drank anything except a potato and a glass of water all day. I never ate meat, fish, or eggs, or drank wine. I was always happy and they never saw me tired, despite the fact that some days I preached as many as twelve sermons.

ANTHONY MARY CLARET

We entered a shop about six times the size of Bocconi's in Milan to buy something we needed and were treated with great kindness and courtesy. We were offered chairs and shown whatever might interest us. In other countries they speak of nobility and courtesy; in London they practice it. This is how they treat Sisters in England, and God, who considers whatever is done to his servants as done to himself, will bless this nation.

FRANCES XAVIER CABRINI

God help both France and Europe. If ever you come to Paris you will be as much struck as I at the dissipation of the place, the never-ending turmoil, the bustle, noise, unrest. How I hate these endless streets which tire my feet, eyes, and ears, where the world and its views reign supreme, and the one object of every living being seems to be pleasure and pleasure only.

BL. THEOPHANE VENARD

If indeed the race of the English—as is noised abroad and is cast up to us in France and Italy—spurn lawful wedlock, a people unworthy and degenerate will be born and the nation will cease to be strong. We suffer for the disgraceful conduct of our people. BONIFACE

Oh brothers, never give up this place of Assisi. If you go anywhere, always return here to your home. FRANCIS OF ASSISI

Now the Greeks leave home and traverse the sea in order to gain an education, but there is no need for us to go abroad on account of the Kingdom of Heaven, nor to cross the sea for virtue. For the Lord has told us before, *the kingdom of God is within you.* ANTHONY OF EGYPT

VIRTUE

Virtue is clearly as problematic as is sin for those the saints must advise; more so, in a way, since the non-saint who becomes zealous about virtue is likely to plunge right into pride, uncharity, hypocrisy, pharisaism at its worst. Scruples (q.v.) are another pitfall for those striving ardently for virtue. Nothing, perhaps, demonstrates the profundity of the doctrine of original sin so clearly as the difficulties we have in the pursuit of virtue.

Through obedience and discipline and training, man, who is created and contingent, grows into the image and likeness of the eternal God.

IRENAEUS

We admire the Creator, not only as the framer of heaven and earth, of sun and ocean, of elephants, camels, horses, oxen, leopards, bears, and lions, but also as the maker of tiny creatures; ants, gnats, flies, worms, and the like—things whose shapes we know better than their names. And as in all creation we reverence His skill, so the mind that is given to Christ is equally earnest in small things as in great, knowing that an account must be given even for an idle word.

JEROME

O religious soul, dove beloved of Christ, behold those little pieces of straw which the world tramples under its feet. They are the virtues practiced by the Savior and thy Spouse, of which He Himself has set thee an example: humility, meekness, poverty, penance, patience, and mortification.

ANTHONY OF PADUA

Nothing richer can be offered to God than a good will; for the good will is the originator of all good and is the mother of all virtues; whosoever begins to have that good will has secured all the help he needs for living well.

ALBERT THE GREAT

Genuine goodness is a matter of habitually acting and responding appropriately in each situation, as it arises, moved always by the desire to please God.   ANONYMOUS, *The Cloud of Unknowing*

Virtue is nothing but well-directed love.   AUGUSTINE OF HIPPO

All the merit of virtue lies in the will. Sometimes a man who wishes to believe has more merit than another who does believe. ALPHONSUS LIGUORI

Virtue demands courage, constant effort, and, above all, help from on high.
                                                           JOHN VIANNEY

Whoever has his mind set on virtue shall be greatly endowed with virtues from above.   BEDE THE VENERABLE

The practice of virtue became attractive, and seemed to come more naturally. At first, my face often betrayed my inward struggle, but little by little sacrifice, even at the first moment, became easier.
                                                     THÉRÈSE DE LISIEUX

Those who wish to live a spiritual life must not fancy that they can make progress in virtue unless they first apply themselves to gaining peace of mind and conscience. Jesus Christ loves to rest in pure and tranquil consciences, a fact which it is easy to understand.   BL. HENRY SUSO

When we pray or meditate on some mystery of the life, passion, and death of our Lord Jesus Christ, we should enter in spirit into the interior of Christ, in order to share in the virtues that He himself practiced in that mystery.
                                                 ANTHONY MARY CLARET

In the case of the virtues, sir, it is very easy to pass from defect to excess, from being just to being rigorous and rashly zealous. It is said that good wine easily turns to vinegar, and that health in the highest degree is a sign of approaching illness.   VINCENT DE PAUL

When, therefore, anyone shall receive the name of abbot, he ought to rule his disciples with a twofold teaching: that is, he should first show them in deeds rather than words all that is good and holy. To such as are understanding, indeed, he may expound the Lord's behests by words; but to the hard-hearted and to the simple-minded he must manifest the divine precepts in his life.   BENEDICT OF NURSIA

My children, the three acts of faith, hope, and charity contain all the happiness of man upon the earth.   JOHN VIANNEY

Don Gonzalo was here not long ago. He is very fond of you, as other people are whom you deluded into having a good opinion of you—though personally I should have been glad if you had been better. Please God you may be better now, and may His Majesty give you the virtue and sanctity that I beseech for you. TERESA OF AVILA, *to her nephew*

CHARITY

Charity is a supernatural virtue and the most important commandment. The saints, who practice it with fervor, are well aware of how unnatural Christian charity is and how different from the love we can easily feel for "good, healthy people who have pleasant manners." Surely, one of the most useful teachings of Christ in helping our human nature stretch itself and in enlarging our imaginative vision is the one cited by Saint Augustine: whatever you do to the most humble and unappealing of my creatures you are doing to Me.

The anecdotes of sainthood are filled with tales of the kissing of lepers and the bringing home of filthy, verminous tramps (as often as not greeted by family or fellow religious with cries of disgust and outrage) who turn out to be Our Lord. Saint Francis did all of the above and is also the most famous of the not insignificant group of saints who extend charity and full communication to the animal kingdom.

Let none of you take a merely natural attitude toward his neighbor, but love one another continually in Jesus Christ.   IGNATIUS OF ANTIOCH

One must see God in everyone.   CATHERINE LABOURÉ

It is easy enough to feel drawn to good, healthy people who have pleasant manners, but that is only natural love and not charity. A mother does not love her sick, deformed child because he is lovable, but because she is his mother, and we must pray the Holy Ghost to put into our hearts that selfless devotion which nature has put into hers.   ROBERT BELLARMINE

I tell you these things, refreshing you, and praying that since we are all created of the same substance, which has a beginning but no end, we may love one another with a single love. For all who know themselves know that they are of one immortal substance.   ANTHONY OF EGYPT

All creatures have the same source as we have. Like us, they derive the life of thought, love, and will from the Creator. Not to hurt our humble brethren is our first duty to them; but to stop there is a complete misapprehension of the intentions of Providence. We have a higher mission. God wishes that we should succor them whenever they require it.   FRANCIS OF ASSISI

That law which is perfect because it takes away all imperfections is charity, and you find it written with a strange beauty when you gaze at Jesus your savior stretched out like a sheet of parchment on the Cross, inscribed with wounds, illustrated in his own loving Blood. Where else, I ask you, my dearest, is there a comparable book of love to read from?
BL. JORDAN OF SAXONY, *to Bl. Diana of Andalo*

If God's word is spoken only naturally, it does very little; but if it is spoken by a priest who is filled with the fire of charity—the fire of love of God and neighbor—it will wound vices, kill sins, convert sinners, and work wonders.   ANTHONY MARY CLARET

The world doth scoff at that which I now say, namely that a man may weep for his neighbor's sins as for his own, or even more than for his own, for it seemeth to be contrary unto nature, but the charity which bringeth this about is not of this world.   BL. ANGELA OF FOLIGNO

If this oath of yours, Mr. Rich, be true, then I pray that I may never see the face of God, which I would not say were it otherwise to win the whole world. In good faith, Mr. Rich, I am sorrier for your perjury than for my own peril.   THOMAS MORE, *at his trial*

Charity never enters a heart without lodging both itself and its train of all the other virtues which it exercises and disciplines as a captain does his soldiers.   FRANCIS DE SALES

Our Savior says that a good tree, that is, a good heart as well as a soul on fire with charity, can do nothing but good and holy works. For this reason Saint Augustine said: "Love and do what you will," namely, possess love and charity and then do what you will. It is as if he had said: "Charity is not able to sin."   ANGELA MERICI

If you truly want to help the soul of your neighbor, you should approach God first with all your heart. Ask him simply to fill you with charity, the greatest of all virtues; with it you can accomplish what you desire.
VINCENT FERRER

If something uncharitable is said in your presence, either speak in favor of the absent, or withdraw, or, if possible, stop the conversation.
JOHN VIANNEY

Do not grieve or complain that you were born in a time when you can no longer see God in the flesh. He did not in fact take this privilege from you. As he says: "Whatever you have done to the least of my brothers, you did to me." AUGUSTINE OF HIPPO

Take scrupulous care never to irritate your husband, your household, or your parents by overmuch churchgoing, exaggerated seclusion, or neglect of your family duties. Don't let it make your censorious of others' conduct, or turn up your nose at conversations which fail to conform to your own lofty standards, for in all such matters charity must rule and enlighten us, so that we comply graciously with our neighbor's wishes in anything that is not contrary to God's law. FRANCIS DE SALES

Humility would despise a good name if charity had no use for it, but because good name is one of the bases of human society and without it we are not only useless but harmful to the public by reason of the scandal it would provoke, charity requires and humility agrees that we should desire to have a good name and carefully preserve it. FRANCIS DE SALES

Peace and union are the most necessary of all things for men who live in common, and nothing serves so well to establish and maintain them as the forebearing charity whereby we put up with one another's defects. There is no one who has not his faults, and who is not in some way a burden to others, whether he be a superior or a subject, an old man or a young, a scholar or a dunce. ROBERT BELLARMINE

POVERTY

Devotion to personal poverty together with love for the poor are central to the saintly vocation. Saint Francis wooed "Lady Poverty" with the intensity of a poet of high chivalry. Love and service to Christ through care of his poor is a very common active apostolate of saints. Materialism which sees nothing positive in poverty can comprehend the second impulse better than the first, understanding of which remains a gift.

Poverty was not found in heaven. It abounded on earth, but man did not know its value. The Son of God, therefore, treasured it and came down from heaven to choose it for Himself, to make it precious to us.

BERNARD OF CLAIRVAUX

The first thing is the love of poverty, whereby the soul putteth away from itself the love of every creature; for it desireth not the possession of any save of the Lord Jesus Christ, it trusteth not in the help of any creature whatsoever in this life; and thus doth love of Him not only reign in the heart but is also shown forth in the works.   BL. ANGELA OF FOLIGNO

The poor monk is lord of the world. He has handed all his cares over to God and by his faith has obtained all men as his servants.   JOHN CLIMACUS

Please remember that you have not joined our Community in order to be comfortable, but to embrace the state of poverty with all its discomforts. You are poor, you say. I am glad of it—it shows that you are happy. You were never as poor as this—so much the better. You never had such opportunities for practicing virtue.

JEAN BAPTISTE DE LA SALLE, *to the Christian Brothers*

The more we despise poverty the more will the world despise us and the greater need will we suffer. But if we embrace Holy Poverty very closely, the world will come to us and will feed us abundantly. FRANCIS OF ASSISI

To the poor in spirit the Kingdom of heaven is assigned as a present recompense, for theirs *is* the Kingdom of Heaven. This is so because to those who are truly poor in spirit the Lord gives great helps, even in this life.

ALPHONSUS LIGUORI

I have been promoted beyond measure by the Lord . . . and I was not worthy nor the kind of person to whom he might grant this since I know for certain that poverty and disaster are more suitable for me than riches and luxury. PATRICK OF IRELAND

If so great and good a Lord, then, on coming into the Virgin's womb, chose to appear despised, needy, and poor in this world, so that people who were in utter poverty and want and in absolute need of heavenly nourishment might become rich in him by possessing the kingdom of heaven, then rejoice and be glad! Be filled with a remarkable happiness and a spiritual joy! Contempt of the world has pleased You more than its honors, poverty more than earthly riches. CLARE OF ASSISI

Love knows that among the poor, and especially among the blind, there are people who shine like the sun, cleansed by their endurance and the ills they have suffered. SIMEON OF EMESA

Even though the poor are often rough and unrefined, we must not judge them from external appearances nor from the mental gifts they seem to have received. On the contrary, if you consider the poor in the light of faith, then you will observe that they are taking the place of the Son of God who chose to be poor. Although in his passion he almost lost the appearance of a man and was considered a fool by the Gentiles and a stumbling block to the Jews, he showed them that his mission was to preach to the poor.

VINCENT DE PAUL

A man's poverty before God is judged by the disposition of his heart, not by his coffers. AUGUSTINE OF HIPPO

If you would rise, shun luxury, for luxury lowers and degrades.

JOHN CHRYSOSTOM

Renunciation of riches is the origin and preserver of virtues. AMBROSE

A detached man should always be looking to see what he can do without.

BL. HENRY SUSO

In the convent we are without shoes and stockings; we shall see if we can stand it. It is certain that on the one hand we do not want to pamper anyone, but on the other hand we do not want to kill anyone either.

VEN. MARY MAGDALEN BENTIVOGLIO

Poverty should be the badge of religious; and as men of the world distinguish their property by stamping it with their names, so the works of religious should be known to be such by the mark of holy poverty.

MARY MAGDELENE DEI PAZZI

Poverty is the true characteristic of a bishop.   ALPHONSUS LIGUORI

The heart of a Christian, who believes and feels, cannot pass by the hardships and deprivations of the poor without helping them.

BL. LOUIS GUANELLA

Take my pocketbook and purse out of my soutane, and if there is any money in them send them to Don Rua. I want to die so poor that they may say that Don Bosco died without leaving a halfpenny.   JOHN BOSCO

Alas! I have nothing to bequeath to you but my bad example.

PAUL OF THE CROSS

My will! What are you talking about? Thank God, I haven't a penny left in the world.   LAWRENCE O'TOOLE

Distribute my garments as follows: let Athanasius, the bishop, have the one sheepskin and the garment I sleep on which he gave me new and which has grown old with me. Let Serapion, the bishop, have the other sheepskin. As to my hair shirt, keep it for yourselves.   ANTHONY OF EGYPT

HUMILITY

Not to be confused with the self-effacement valued in those eastern religions that seek release from the prison of the ego, true humility is perhaps the most distinctively Christian of all the virtues. It is not simply the outward manifestation of the realization of one's unworthiness before God, but an imitation of the example of Christ's incarnation and humiliation. God has personified this virtue. For this reason, humility has a transcendent quality; it is a positive, brilliant, divine attribute, to be sought and embraced with ardent vigor. It is also, explicitly, the way to triumph. "Whosoever shall exalt himself shall be abased; and he that shall humble himself shall be exalted" (Matthew 23:12).

It seems to me that humility is truth. I do not know whether I am humble, but I do know that I see the truth in all things.　　THÉRÈSE DE LISIEUX

Wouldst thou comprehend the height of God? First comprehend the lowliness of God. Condescend to be humble for thine own sake, seeing that God condescended to be humble for thy sake too, for it was not for his own.
　　AUGUSTINE OF HIPPO

It was pride that caused the fall of Lucifer and Adam. If you should ask me what are the ways of God, I would tell you that the first is humility, the second is humility, and the third is still humility. Not that there are no other precepts to give, but if humility does not precede all that we do, our efforts are fruitless.　　AUGUSTINE OF HIPPO

Whoever will proudly dispute and contradict will always stand outside the door. Christ, the master of humility, manifests His truth only to the humble and hides Himself from the proud.　　VINCENT FERRER

A man is humble when he stands in the truth with a knowledge and appreciation for himself as he really is.　　ANONYMOUS, *The Cloud of Unknowing*

Blessed is the servant who esteems himself no better when he is praised and exalted by people than when he is considered worthless, simple, and despicable; for what a man is before God, that he is and nothing more.

FRANCIS OF ASSISI

Christ is a sun of righteousness and also of mercy, Who stands in the highest part of the firmament, on the right hand of the Father, and from thence He shines into the bottom of the humble heart; for Christ is always moved by helplessness whenever a man complains of it and lays it before Him with humility.   BL. JAN VAN RUYSBROECK

Humility and self-contempt will obtain our wish far sooner than stubborn pride. Though God is so exalted, His eyes regard the lowly, both in heaven and earth, and we shall strive in vain to please Him in any other way than by abasing ourselves.   JOHN OF AVILA

But I beg those who believe in God and fear him whoever shall condescend to peruse or to receive this writing which Patrick, a very badly educated sinner, has written in Ireland, that nobody shall ever say that it was I, the ignoramus, if I have achieved or shown any small success according to God's pleasure, but you are to think, and it must be sincerely believed, that it was the gift of God.   PATRICK OF IRELAND

I knew nothing; I was nothing. For this reason God picked me out.

CATHERINE LABOURÉ

In order to banish this plague of vainglory far from my children, I do ask you and their mother and all their friends to sing this song to them, and go on singing it, drive it into their heads that vainglory is despicable and to be spit upon, and that there is nothing more sublime than the humble modesty so often praised by Christ.   THOMAS MORE

Humility must always be doing its work like a bee making its honey in the hive: without humility all will be lost.   TERESA OF AVILA

A certain degree of splendor is necessary if the dignity of the sacred Order is to receive its due meed of respect from the world at large. I am trying as hard as ever I can to keep *my* splendor and dignity as modest as may be. Among those of my colleagues who are neither extravagant nor showy, but follow a middle course that has, however, its own elegance and distinction, I hold the least elegant and distinguished place. Indeed, within the limits of decorum and dignity, I am just not shabby.   ROBERT BELLARMINE

Humility is like a pair of scales; the lower one side falls, the higher rises the other. Let us humble ourselves like the Blessed Virgin and we shall be exalted.   JOHN VIANNEY

This lowliness of heart is mother of all the virtues, whence springeth even the exercising of these virtues, as the trunk and branches spring from the root. So precious is this virtue of humility and so firm its foundation (upon which is built up the whole perfection of the spiritual life), that the Lord did especially desire that we learn it direct of him. BL. ANGELA OF FOLIGNO

Often, actually very often, God allows His greatest servants, those who are far advanced in grace, to make the most humiliating mistakes. This humbles them in their own eyes and in the eyes of their fellow men. It prevents them from seeing and taking pride in the graces God bestows on them or in the good deeds they do, so that, as the Holy Ghost declares: "No flesh should glory in the sight of God." LOUIS-MARIE GRIGNION DE MONTFORT

True humility does not make a show of itself and hardly speaks in a humble way. It not only wants to conceal all other virtues but most of all it wants to conceal itself. If it were lawful to lie, dissemble, or scandalize our neighbor, humility would perform arrogant, haughty actions so that it might be concealed beneath them and live completely hidden and unknown.

FRANCIS DE SALES

Not to feel pleased at being praised is, I am afraid, what has never happened to any man. JOHN CHRYSOSTOM

God can make use of all sorts of ways to show his will to man. Sometimes he makes use of the most unfitted and unworthy instruments, like Balaam's ass whom he caused to speak, or Balaam himself, false prophet as he was, who foretold many things concerning the Messiah. So may it be with me.

JOHN BOSCO

In order to become an instrument in God's hands we must be of no account in our own eyes. ANGELA MERICI

There was a man who struck Socrates full in the face, dealing him blow upon blow. Socrates offered no resistance; he allowed the drunkard's anger free rein. Within a short time his face was swollen and livid from bruises. When at last his assailant gave over, Socrates simply wrote on his forehead as an artist engraves his name on a statue: "So and So did this." That was his sole retaliation. Surely it is well for us in our generation to imitate such models. BASIL THE GREAT

The spiritually self-sufficient do often fall into error and are more difficult to correct than those who have worldly self-sufficiency. Esteem yourselves, therefore, as nothing, as nothing known or nothing unknown. Of a truth, the soul can possess no better insight or knowledge than to perceive its own nothingness and to remain within its own prison. BL. ANGELA OF FOLIGNO

The very moment God sees us fully convinced of our nothingness, He reaches out His hand to us.   THÉRÈSE DE LISIEUX

OBEDIENCE

The saints give obedience its due as the great training tool of all virtues: it requires the mortification of self-will, of pride, and of a host of forms of self-deception. It is the most demanding and the crown of virtues, magnificent in its holy irrationality. Wonderful stories are told of medieval feats of obedience, which have a paradoxical, Zen-like quality, as in the case of the monk who at the command of his superior faithfully watered a dry stick for three years to find it eventually blossoming into a miraculous tree.

Eichmann-like perversions of "obedience" have perhaps made us unduly wary of this virtue. Obedience to God is without limit, whereas obedience to human beings and institutions is limited by higher laws that must not be transgressed. Although somewhat out of favor in the contemporary Church, as well as in ordinary society, obedience is glorious—much of the time— "because it is absurd."

In the first Adam we offended God by not performing his command; in the second Adam we have been reconciled, becoming "obedient unto death."

IRENAEUS

Christ was subject unto his most poor and humble Mother and His putative father, obedient unto them and humbly serving until his thirtieth year. He was obedient in the midst of His disciples, who were few in number, ignorant, and poor. He said He was not come to be ministered unto but to minister unto them. BL. ANGELA OF FOLIGNO

Obedience is the only virtue that implants the other virtues in the heart and preserves them after they have been so implanted. GREGORY I THE GREAT

Obedience is the perfection of the religious life; by it man submits to man for the love of God, as God rendered Himself obedient unto men for their salvation. THOMAS AQUINAS

Obedience is the mortification of the members while the mind remains alive. Obedience is unquestioned movement, death freely accepted, a simple life, danger faced without worry, an unprepared defense before God, fearlessness before death, a safe voyage, a sleeper's journey. Obedience is the sepulcher of the will and the resurrection of lowliness.   JOHN CLIMACUS

Isn't it extraordinary, Mother, what a lot of nervous strain you can avoid by taking the vow of obedience? How enviable it is, the simple creed of the religious, who has only one compass to steer by, the will of her superiors! She knows for certain, all the time, that she is on the right path: there's no fear that she can go wrong, even when she feels fairly certain that her superiors are wrong.   THÉRÈSE DE LISIEUX

You should accept as a grace all those things that deter you from loving the Lord God and whoever has become an impediment to you, whether they are brothers or sisters, even if they lay hands on you. And you should desire that things be this way and not otherwise. And let this be an expression of true obedience to the Lord God.   FRANCIS OF ASSISI

If you are ordered to eat meat, be extremely careful not to make the slightest resistance to obedience, for you will render more honor to God by eating flesh meat through obedience than by fasting on bread and water of your own volition.   JEAN EUDES

He who wishes to make an absolutely complete offering of himself must in addition to his will include his understanding, which is the third and highest degree of obedience. He not only identifies his will with that of his superior, but even his thought.   IGNATIUS OF LOYOLA

If you begin to grieve at this, to judge your superior, to murmur in your heart, even though you outwardly fulfill what is commanded, this is not the virtue of obedience, but a cloak over your malice.

BERNARD OF CLAIRVAUX

The virtue of obedience makes the will supple. It gives the power to conquer self, to overcome laziness, and to resist temptations. It inspires the courage with which to fulfill the most difficult tasks.   JOHN VIANNEY

Obedience is a little dog that leads the blind.   JOSEPH OF COPERTINO

Thus the abbot John, without a single thought as to whether it would do any good or not, with great and prolonged labor watered a dry stick for a whole year on end when told to do so. Heaven sometimes approved this kind of obedience with miracles.   IGNATIUS OF LOYOLA

In other sacrifices the flesh of another is slain, but in obedience our own will is sacrificed.   GREGORY I THE GREAT

Obedience is a whole burnt-offering in which the entire man, without the slightest reserve, is offered in the fire of charity to his Creator and Lord by the hands of His ministers.   IGNATIUS OF LOYOLA

LOVE

Together with the immoderate love of God, Christianity also is capable of urging a chilling detachment from the claims of natural loves, especially those of family. "He that loveth father or mother more than me is not worthy of me; and he that loveth son or daughter more than me is not worthy of me" (Matthew 10:37). The saints speak of love of God and love of creatures as being to some degree opposed or in conflict. Limited as we are, it seems to be hard to do both at once; most of our saints emphasize love of God or love of specific creatures at any one time.

The most intensely passionate natures, like those of Thérèse and Ignatius, are perhaps more drawn toward God as the only sufficient object than are gentler saints like Anselm or more versatile natures like that of Bernard. Great mystics do have tender worldly loves, but often they take the form of intense family attachments of a very domestic type. It is touchingly evident, however, that the saints often loved each other dearly and even "fell in love."

Mystics remind us that God is the greatest lover of all. Being chosen for a direct experience of this overwhelming love can be intoxicating, or a kind of madness. Teresa of Avila once noted that she had "been going around as if drunk."

Perhaps the most distinctively Christian form of love springs from the commandment to love those who are not lovable. "Love your enemies, do good to them which hate you, bless them that curse you, and pray for them which despitefully use you" (Luke 6:27).

We were born to love, we live to love, and we will die to love still more.

JOSEPH CAFASSO

God did not make the first human because He needed company, but because He wanted someone to whom he could show His generosity and love. God did not tell us to follow Him because He needed our help, but because He knew that loving Him would make us whole.   IRENAEUS

The soul cannot live without love. All depends on providing it with a worthy object.   FRANCIS DE SALES

Though we cannot know God, we can love him: by love he may be touched and embraced, never by thought.   ANONYMOUS, *The Cloud of Unknowing*

Lord, I dare not say I love Thee, but I *will* love Thee.
                                           PETER JULIAN EYMARD

At last I have found my calling! My calling is love!   THÉSÈSE DE LISIEUX

The way of salvation is easy; it is enough to love.   MARGARET OF CORTONA

How fine it would be, father, if one day you could say, "Gemma was love's victim, and it was purely of love that she died."   GEMMA GALGANI

To love God you need three hearts in one—a heart of fire for Him, a heart of flesh for your neighbor, and a heart of bronze for yourself.
                                           BENEDICT JOSEPH LABRE

Love Him totally Who gave Himself totally for your love.   CLARE OF ASSISI

I shall always be grateful to our Lord for turning earthly friendships into bitterness for me, because, with a nature like mine, I could so easily have fallen into a snare and had my wings clipped; and then how should I have been able to "fly away and find rest"? I don't see how it's possible for a heart given over to such earthly affections to attain any intimate union with God.   THÉRÈSE DE LISIEUX

The truth is I can love a person in this life only so far as he strives to advance in the service and praise of God our Lord; for he who loves anything for itself and not for God, does not love God with his whole heart.
                                           IGNATIUS OF LOYOLA

The important thing is not to love Me for your own sake, or yourself for your own sake, or your neighbor for your own sake, but to love Me for Myself, yourself for Myself, and your neighbor for Myself. Divine love cannot suffer to share with any earthly love.   CATHERINE OF SIENA

Instead of the kiss which I am prevented from giving you, I am sending, by the bearer of this letter, two little kegs of wine. As you love me, I beg you to use it for a day of rejoicing with your friends.

BONIFACE, *to Egbert, archbishop of York*

O you good man, what have you done? You have praised a sinner, you have numbered a good-for nothing amongst the blessed! You must now pray that I shall not be led into error if, in my delight at such great praises, I were to forget the sort of person I am. This almost did happen when I read the letter in which you made me out to be blessed. If words could do that, how happy I would be. Even now I would call myself happy, but by your favor, not by my own deserts. Happy to be loved by you and happy in loving you. Would that I could enjoy your company, I do not say always, nor even often, but just once or twice in a year.

BERNARD OF CLAIRVAUX, *to Peter of Cluny*

You can't imagine how much I love you when I see you covered with persecutions! If only someone could arrange for me to see you and let me put my arms about your neck, the way a little boy hangs on to his dear father.

ANTHONY MARY CLARET, *to the Bishop of Palencia*

Souls well-beloved of my soul, my eyes ardently desire to behold you; my arms expand to embrace you; my lips sigh for your kisses; all the life that remains to me is consumed with waiting for you. How can I forget those whom I have placed like a seal upon my heart?

ANSELM OF CANTERBURY, *to his family*

It is only a small thing that we write to each other; there is a fire of love in our hearts, in the Lord, and there you speak to me and I speak to you the whole time, in feelings of affection which no tongue could adequately express and which no letter could adequately contain.

BL. JORDAN OF SAXONY, *to Bl. Diana of Andalo*

When you receive a letter from a friend, dearest son, you should not delay to embrace it as a friend. For it is a fine consolation among the absent that if one who is loved is not present, a letter may be embraced instead.

ISIDORE OF SEVILLE

The more we know of men, the less we love them. It is the contrary with God; the more we know of him, the more we love him.   JOHN VIANNEY

In regard to all persons, have the same love, have the same indifference, whether relations or strangers. Detach your heart as much from the one as from the other; in a sense, even more particularly from relations, for fear lest flesh and blood be stirred with the normal love which must be forever mortified if one is to achieve spiritual perfection.   JOHN OF THE CROSS

There is neither father nor mother nor son, nor any other person whatsoever, who can embrace the object beloved with so great a love as that wherewith God embraceth the soul.   BL. ANGELA OF FOLIGNO

For we shall see verily in Heaven, without end, that we have grievously sinned in this life, and notwithstanding this, we shall see that we were never hurt in His love, nor were less of price in His sight. For hard and marvelous is that love which may not, nor will not be broken for trespass.

BL. JULIAN OF NORWICH

The inward stirring and touching of God makes us hungry and yearning; for the Spirit of God hunts our spirit; and the more it touches it, the greater our hunger and our craving. And this is the life of love in its highest working, above reason and above understanding; for reason can here neither give nor take away from love, for our love is touched by the Divine love.

BL. JAN VAN RUYSBROECK

The Knight of the Immaculate does not confine his heart to himself, nor to his family, relatives, neighbors, friends, or countrymen, but embraces the whole world, each and every soul, because, without exception, they have all been redeemed by the blood of Jesus. They are all our brothers. He desires true happiness for everyone, enlightenment in faith, cleansing from sin, inflaming of their hearts with love toward God and love toward neighbor, without restriction.   MAXIMILIAN KOLBE

My brothers, Christ made love the stairway that would enable all Christians to climb to heaven. Hold fast to it, therefore, in all sincerity, give one another practical proof of it, and by your progress in it, make your ascent together.   FULGENTIUS OF RUSPE

And I, dear friend, am torn by the desire to see you, so that I may receive a dying man's blessing. I who have loved you from the first shall love you without end. I say with all confidence that I can never lose one whom I have loved unto the end; one to whom my soul cleaves so firmly that it can never be separated does not go away but only goes before. Be mindful of me when you come to where I shall follow you.

BERNARD OF CLAIRVAUX, *to Abbot Suger*

Love for our neighbor consists of three things: to desire the greater good of everyone; to do what good we can *when* we can; to bear, excuse, and hide other's faults.   JOHN VIANNEY

The interior life is like a sea of love in which the soul is plunged and is, as it were, drowned in love. Just as a mother holds her child's face in her hands to cover it with kisses, so does God hold the devout man.

JOHN VIANNEY

FORGIVENESS

Forgiveness springs from humility, in the realization of our own weakness, imperfection, and sinfulness. If God, who is perfect, can forgive sinners, how much more should humans do so. "Judge not, and ye shall not be judged: condemn not, and ye shall not be condemned: forgive, and ye shall be forgiven" (Luke 6:37).

It is perhaps easier to rise to a Christ-like level of forgiveness of one's murderers, executioners, or torturers, once and for all, than it is to forgive those who try us in small ways, every day. Forgiveness within the family, forgiving those whom we have loved and trusted, is a stringent test.

I beg you for pity's sake to show a little more charity toward your son. He came to see me this morning at Naples and burst into tears in front of me. . . . Remember that he is your son and not some mongrel or other, and surely far dearer as such than property or wealth. You may be certain of this: if you practice love within your own family, God will help you outside it. ALPHONSUS LIGUORI, *to his father*

The sooner you forgive him, the sooner he will recover from his illness. You are the person to restore him to health of soul and body. Speak to your son again. Do not refuse me, Father! If I am truly your daughter and you love me as much as you profess, you will grant me what I ask.

CATHERINE DEI RICCI

Tell me, how are we two going to face the Day of Judgement? The sun is witness that it has gone down on our anger not one day but for many a long year. . . . Woe to me, vile wretch! Must I say, Woe to you also? . . . I am now renewing the request I made to you in my previous letter of a year ago. . . . Soon harmony restored or harmony ruptured will receive reward or penalty before His tribunal. Very well, if you now rebuff me, the guilt will not be on my head. Once you have read it, this letter of mine will secure my acquittal. JEROME, *to his aunt*

A man who had resolved to take vengeance on another will change his mind in the confessional, but a little later you will find him among his friends talking delightedly about his quarrel and saying, "If it wasn't for the fear of God, I would do this or that," "In this matter of forgiving people the divine law is a hard thing," and "I wish to God it would let a man revenge himself."   FRANCIS DE SALES

The saints had no hatred, no bitterness. They forgive everything and think they deserve much more for their offenses against God.   JOHN VIANNEY

Pardon one another so that later on you will not remember the injury. The recollection of an injury is in itself wrong. It adds to our anger, nurtures our sin and hates what is good. It is a rusty arrow and poison for the soul. It puts all virtue to flight.   FRANCIS OF PAOLA

We should love and feel compassion for those who oppose us, rather than abhor and despise them, since they harm themselves and do us good, and adorn us with crowns of everlasting glory while they incite God's anger against themselves.   ANTHONY ZACCARIA

September 12, 1869. At 11:30 in the morning, the Lord granted me the grace of love for my enemies. I felt it in my heart. ANTHONY MARY CLARET

I for the love of Jesus forgive my murderer and I want him to be with me in paradise. May God forgive him, because I have already forgiven him.
                                                            MARIA GORETTI

I forgive my capital persecutor who hath been so long craving and thirsting after my blood; from my soul I forgive him, and wish his soul so well that were it in my power I would seat him a seraph in heaven.   DAVID LEWIS

Would God I might suffer ten times as much that thou might go free for the blow thou hast given me. I forgive thee, and pray to God to forgive thee even as I would be forgiven.   BL. THOMAS WOODHOUSE

No force can prevail with a Father like the tears of his child, nor is there anything which so moves God to grant us, not justice, but mercy, as our sorrow and self-accusation.   JOHN OF AVILA

SATAN AND
HIS LEGIONS

———

Satan, "the adversary," is the brilliant chief of the fallen angels, exiled from heaven because of his envy and pride. He and his demons are a personal, malign force at loose in the world which seeks to thwart the designs of God by tempting men. "Put on the whole armour of God that ye may be able to stand against the wiles of the devil. For we wrestle not against flesh and blood but against principalities, against powers, against the rulers of the darkness of this world" (Ephesians 6:11–12).

Satan is also identified by Jesus as "the Prince of this world" (John 14:30), a murderer, and the father of lies (John 8:44). Tempting Christ in the wilderness (Matthew 4:1–11), Satan claims to have the world in his gift and is not contradicted. He is a formidable opponent who knows Scripture well and quotes it aptly.

Satanists view the principle of evil as sharing equally with good in ruling the universe. Thomas Aquinas, expressing the orthodox Christian view, says that good is the only reality and that evil is a contingent deprivation. Satan as a person, a fallen angel, expresses this situation perfectly. Satan was created by God, banished by God, and holds his influence by God's sufferance.

The saints are confident that God loves and controls everything, and they are not daunted by the existence of apparent evils that cause such difficulty in the thinking of unbelievers who mock the idea of a just God who can allow suffering to exist.

———

Modern times are dominated by Satan and will be more so in the future. The conflict with hell cannot be engaged by men, not even the most clever.

MAXIMILIAN KOLBE

Now, schemes for working evil come easily to the devil, so when it was nighttime they made such a crashing noise that the whole place seemed to be shaken by a quake. The demons, as if breaking through the building's four walls, and seeming to enter through them, were changed into the forms of beasts and reptiles. The place immediately was filled with the appearances of lions, bears, leopards, bulls, and serpents, asps, scorpions, and wolves.   ATHANASIUS, *The Life of Anthony*

Demons are very cowardly, always anxious about the fire that has been prepared for them. To bolster your courage against them, take this sure sign. When some apparition occurs, do not collapse in terror, but whatever it may be ask first, bravely, "Who are you, and where do you come from?" If it is a divine vision, they will give you assurances and change your fear to joy. If it is some devil, it will immediately be weakened by your formidable spirit.   ANTHONY OF EGYPT

Be eager for more frequent gatherings for thanksgiving to God and his glory, for when you meet thus, the forces of Satan are annulled and his destructive power is cancelled in the concord of your faith. IGNATIUS OF ANTIOCH

It was the ground that God cursed, not Adam. And he cursed the serpent; and the fire was prepared originally for the devil and his angels, not for man.   IRENAEUS

Christ did give the devil power over Him that He might be tempted and led into danger and persecuted even unto death, in order that He might thereby liberate man from the devil's power.   BL. ANGELA OF FOLIGNO

Our Lord's body became a bait for death so that the dragon, hoping to swallow him, might vomit up also those whom he had swallowed.
                                      CYRIL OF JERUSALEM

The Lord called his disciples the salt of the earth because they seasoned with heavenly wisdom the hearts of men, rendered insipid by the devil.
                                      CHROMATIUS OF AQUILEIA

To free a man who is bodily a captive in the hands of barbarians is a noble deed, but to free a soul from the slavery of Satan is greater than to deliver all who are in corporal slavery.   JEAN EUDES

Let us rush with joy and trepidation to the noble contest and with no fear of our enemies. They are invisible themselves, but they can see the condition of our soul. If they see our spirits cowering and trembling, they will make a more vigorous attack against us. Let us arm ourselves against them with courage. They hesitate to grapple with a bold fighter. JOHN CLIMACUS

You damned spirits! You can only do what the hand of God allows you to do. Therefore in the name of Almighty God I tell you to do whatever God allows you to do to my body. I will gladly endure it since I have no worse enemy than my body. If you take revenge on my enemy for me, you do me a very great favor. FRANCIS OF ASSISI

The way to overcome the Devil when he excites feelings of hatred for those who injure us is immediately to pray for their conversion. JOHN VIANNEY

In the case of those who are making progress from good to better the good angel touches the soul gently, lightly, sweetly, as a drop of water enters a sponge, while the evil spirit touches it sharply, with noise and disturbance, like a drop of water falling on a rock. In the case of those who go from bad to worse, the contrary happens, and the reason for this is the disposition of the soul. When the soul is contrary to these spirits they enter with perceptible commotion, but when it is similar to them they enter in silence, as into their own house, through open doors. IGNATIUS OF LOYOLA

Every evil is based on some good. Indeed, evil cannot exist by itself, since it has no essence, as we have shown. Therefore, evil must be in some subject. Now, every subject, because it is some sort of substance, is a good of some kind. So, every evil is in a good thing. THOMAS AQUINAS

The demons were not created as the figures we now identify as 'demonic', for God made nothing bad. They were made good, but falling from the heavenly wisdom and thereafter wandering around the earth, they deceived the Greeks through apparitions. Envious of us Christians, they meddle with all things in their desire to frustrate our journey into heaven so that we might not ascend to the place from which they fell. ANTHONY OF EGYPT

Heed not the demons even if they awaken you for prayer, or counsel you to eat nothing at all, or level accusations and reproaches for actions for which, at some other time, they urged excuses on you. They do these things not for the sake of piety or truth, but to bring the simple to despair and to declare the discipline useless. They want to make men sick of the solitary life. ANTHONY OF EGYPT

Consider that the devil does not sleep, but seeks our ruin in a thousand ways. ANGELA MERICI

First tell the Devil to rest, and then I'll rest too. JOHN BOSCO

The fiend selects for his most furious attacks the time when we feel most unable to pray. NILUS SORSKY

That the soul may advance with sure steps in spiritual ways, it must walk with constancy in the opposite direction from that in which the enemy of salvation wishes to lead it. If he seeks to relax the conscience, let it contract. If he seeks to contract it, let it relax. Avoiding the two extremes, it will establish itself in a middle path that will be for it a state of assurance and peace. IGNATIUS OF LOYOLA

What joy you give me when you tell me that you continue to go to Holy Communion. But I do not feel happy about the rest of your letter when you speak of the silly fears which you are allowing to grip you once again. All your apprehensions spring from the Devil and they will lead you, I am afraid, to give up your communions. You tremble for your past? I tremble for your future! ALPHONSUS LIGUORI

Believe me when I say that the devil has his contemplatives as surely as the Lord has his. ANONYMOUS, *The Cloud of Unknowing*

SIN

Sin, says St. Augustine, is "a word, deed, or desire in opposition to the eternal law." Sin is present whenever man tries to separate himself from God, ceases to acknowledge his dependence on God, and refuses God's gifts. Pride is the primal, fundamental, original sin, which denies the faithfulness of God and leads to disobedience to the divine will.

The Church has always taught that all sins, no matter how grave, can be forgiven. Sins that prevent the acceptance of grace and repentance present special problems, however. These "sins against the Holy Spirit" are despair, presumption, obstinacy in sin, envy of another's sanctity, rejection of known truths of faith, and impenitence.

The specific laws of God that are to be kept include the ten commandments of the Mosaic law and the two "great commandments" of Christ: "Thou shalt love the Lord thy God with all thy heart, and with all thy soul, and with all thy mind. This is the first and great commandment. And the second is like unto it, Thou shalt love thy neighbor as thyself" (Matthew 22:37–39). Failure to follow these count as sins against faith and charity.

The capital or cardinal sins are so called because they are vices, passions, or habits that lead to a host of actual sins. These are pride (found in all sin to some degree, whose opposing capital virtue is humility), envy (opposed to charity), covetousness (opposed to liberality and justice), lust (opposed to chastity), anger (opposed to meekness), gluttony (opposed to temperance), and sloth (opposed to charity).

We have only one evil to fear and that is sin.   ALPHONSUS LIGUORI

Enjoy yourself as much as you like—if only you keep from sin.

JOHN BOSCO

See, you blind ones, you who are deceived by your enemies; by the flesh, the world, and the devil; because it is sweet to the body to commit sin and it is bitter for it to serve God.   FRANCIS OF ASSISI

So did my two wills, one new, the other old, one spiritual, the other carnal, fight within me and by their discord undo my soul.   AUGUSTINE OF HIPPO

This foul, wretched lump called sin is none other than yourself and though you do not consider it in detail, you understand now that it is part and parcel of your very being and something that separates you from God.
                                 ANONYMOUS, *The Cloud of Unknowing*

There is only one thing to be feared, Olympias, only one trial, and that is sin. I have told you this over and over again. All the rest is beside the point, whether you talk of plots, feuds, betrayals, slanders, abuses, accusations, confiscations of property, exile, sharpened swords, open sea, or universal war. Whatever they may be, they are all fugitive and perishable. They touch the mortal body but wreak no harm on the watchful soul.
                                                   JOHN CHRYSOSTOM

Sin is a cruel murder, a frightful act of deicide, a ghastly annihilation of all things. It is murder because it is the only cause of death, both of the body and the soul of man. It is deicide because sin and the sinner caused Christ to die on the Cross and the sinner continues this crucifixion of Jesus. day by day, within himself.   JEAN EUDES

I do not say that sin should not be displeasing unto you and that you should not hold it in the greatest horror, but I do say ye should never judge the sinners, and moreover I say that ye should never despise them, for ye know not the judgements of the Lord God.   BL. ANGELA OF FOLIGNO

Unless the will is set and deliberate there is no sin; neither is there if reason opposes temptation, and if the evil thoughts which present themselves are repugnant to it. There is sin only when, with the full knowledge and by the fully consenting act of the will, without hesitation or repugnance, the soul yields itself to the evil suggestion.   BL. HENRY SUSO

Weak, lazy penitents abstain regretfully for a while from sin. They would very much like to commit sins if they could do so without being damned. They speak about sin with a certain petulance and with a liking for it and think those who commit sins are at peace with themselves.
                                                   FRANCIS DE SALES

One must deal with some sinners as with snails: put them first in cool water until they come out, and then cook them little by little before they realize what's happening to them.   ANTHONY MARY CLARET

The contemplative work of love by itself will eventually heal you of all the roots of sin. Fast as much as you like, watch far into the night, rise long before dawn, discipline your body, and if it were permitted—which it is not—put out your eyes, tear out your tongue, plug up your ears and nose, and cut off your limbs; yes, chastise your body with every discipline and you would still gain nothing. The desire and tendency toward sin would remain in your heart.   ANONYMOUS, *The Cloud of Unknowing*

You have had the audacity to try and dissuade a soldier of Christ from the service of his Lord. I tell you, there is one who will see and judge this. Are not your own sins enough for you that you must saddle yourself with the sins of another by doing your best to entice a repentant young man back to his follies and thus, in your hard and unrepentant heart to lay up wrath for yourself on That Day? As though the devil were not tempting Peter enough without the help of you who are supposed to be a Christian and his friend and leader!   BERNARD OF CLAIRVAUX

God pitied man and sought to prevent him from continuing in sin forever and in evil without end or remedy, so He put a stop to his wickedness by interposing death and thus making his sin to cease, so that man by dying to sin should begin to live to God.   IRENAEUS

Keep yourself, my son, from everything that you know displeases God, that is, from every mortal sin. You should permit yourself to be tormented by every kind of martyrdom before you would allow yourself to commit a mortal sin.   LOUIS IX OF FRANCE, *to his son*

Death, but not sin!   DOMINIC SAVIO

Have courage, then, and conquer by your generosity the shame that the devil magnifies so much in your mind. It will be enough to begin to reveal the sin you have committed; all your ridiculous apprehensions will vanish on the spot. And, believe me when I tell you that afterwards you will feel more happy at having confessed your sins than if you had been made monarch of the whole earth.   ALPHONSUS LIGUORI

No mother could snatch her child from a burning building more swiftly than God is constrained to succor a penitent soul, even though it should have committed every sin in the world a thousand times over.

BL. HENRY SUSO

Just as water extinguishes a fire, so love wipes away sin.   JOHN OF GOD

TEMPTATIONS

Temptation means "putting to the test" and describes any enticement to sin. God permits us to be tempted to make us realize our weakness, to test our faith, and to help us by his grace to strengthen virtue by practice. Sources of temptation to sin are the attractiveness of bad example in the world, the urges and lusts of the flesh, and the evil spirits who encourage all forms of pride and covetousness.

Man is vulnerable to temptation because he lost the gifts of innocence, understanding, wisdom, and counsel with the original sin and is thus heir to death, suffering, ignorance, and a strong inclination to error and to sin.

God himself may test his servants, as in the case of Abraham, or give permission for Satan to do so, as in the story of Job. Conversely, man can tempt God in challenging him to prove his power; this is expressly forbidden as a sin against faith, as Jesus reminds Satan during the temptation in the desert.

Jesus told his disciples to "Watch and pray, that ye enter not into temptation. The spirit indeed is willing, but the flesh is weak" (Matthew 26:41). Both prudence and the help of grace are necessary to avoid and overcome temptation.

Hermits like St. Anthony have experienced the most vivid and clearly satanic temptations in the form of hallucinations. Satan may not need to put on any special show for people in the world, where natural temptations abound and need only be taken advantage of, but in the desert he creates erotic or terrifying visions to catch the attention of the holy hermits.

Alas, my God, I know that thou hast made all things well and that by Thy grace I shall not sin in these temptations, but I would wish not to experience them.   JOSEPH OF COPERTINO

The greatest of all evils is *not* to be tempted, because there are then grounds for believing that the devil looks upon us as his property.   JOHN VIANNEY

I consider you more of a servant and friend of God and I love you more, the more you are attacked by temptations. Truly I tell you that no one should consider himself a perfect friend of God until he has passed through many temptations and tribulations.   FRANCIS OF ASSISI

Let us be clear about this: the fiend must be taken into account. Anyone beginning this work of contemplation (I do not care who he is) is liable to feel, smell, taste or hear some surprising effects concocted by this enemy in one or another of his senses. So do not be surprised if it happens. There is nothing he will not try in order to drag you down from the heights of such valuable work   ANONYMOUS, *The Cloud of Unknowing*

Once while I was fasting, the cunning one even came as a monk, carrying what looked like loaves of bread and he counselled me, saying, "Eat and rest from your labors. You are only a man and you will soon grow weak." Perceiving his strategy, I rose to say my prayers instead, and he could not endure this, so he fled like wisps of smoke that pass through the doorway.
                              ANTHONY OF EGYPT

Here was one lady talking about my pretty hair and another, just going out the door and thinking I couldn't hear, wanting to know who that very pretty girl was. Compliments are at their best when you aren't meant to hear them and the thrill of pleasure I felt made me realize that I was full of self-love. I'm always ready to sympathize with the people who lose their souls—after all, it's so easy, once you begin to stray along the primrose path of worldliness.   THÉRÈSE DE LISIEUX

I do not trust myself as long as I am in this body of death because he is strong who daily strives to seduce me from my faith and the purity of a sincere religion. . . . But the hostile flesh always draws me towards death, that is towards enticements unlawful to indulge in.   PATRICK OF IRELAND

He said not: Thou shalt not be troubled—thou shalt not be tempted—thou shalt not be distressed. But He said: Thou shalt not be overcome.
                              BL. JULIAN OF NORWICH

Close your ears to the whisperings of hell and bravely oppose its onslaughts.   CLARE OF ASSISI

I wish I could fight in your stead, receiving in my own soul the attacks and the wounds that you are enduring. But if I could do that then you would not receive in Heaven, with Christ's other soldiers, the palm of victory.

BL. HENRY SUSO

(1) The thought comes to me to commit a mortal sin. I resist the thought immediately and it is conquered. (2) If the same evil thought comes to me and I resist it and it returns again and again, but I continue to resist it until it is vanquished. This second way is much more meritorious than the first.

IGNATIUS OF LOYOLA

A man who governs his passions is master of the world. We must either command them or be enslaved to them. It is better to be a hammer than an anvil.   DOMINIC

No one ought to be confident in his own strength when he undergoes temptation. For whenever we endure evils courageously, our long-suffering comes from Christ.   AUGUSTINE OF HIPPO

There is one sort of temperance for those of good conduct and another for those inclined to particular weaknesses. Among the former any kind of bodily stirring evokes an immediate urge to restraint, while among the latter there is no relief or relaxation from such stirrings until the very day they die. The former strive always for peace of mind, while the latter try to appease God by their contrition.   JOHN CLIMACUS

If I could be born again, I would have fewer desires.   FRANCIS DE SALES

SCRUPLES

A scruple is morbid anxiety about the possible sinfulness of an action or unreasonable worry that a trivial matter is really one of moral gravity. Scruples can reflect obsessive neurosis, confusion, poor judgement, an overly delicate conscience, ill health, or anxiety states. Remedies are prayer, trust in God, restoration of health, and obedience to a spiritual counselor.

Many saints have suffered from scruples at some point in their development. Only the soldierly, tough-minded Ignatius of Loyola has a good word to say about them; to most saints they are a serious nuisance and torment, exasperating to confessors, and maddening for the sufferers.

Victims of scruples used to flock from all over Europe to consult famous confessors like Cafasso of Turin or Vianney of Ars, just as more recent generations seeking relief from compulsive neuroses traveled to Vienna or Zurich.

All your intellectual and moral faculties tell you that you have not sinned, yet in your conscience the idea arises that you have done wrong. From this comes perplexity and trouble, which the evil spirit keeps up; this is a scruple, properly speaking.   IGNATIUS OF LOYOLA

Scrupulous souls, forever tormented by doubts and anxiety have hearts which are ill prepared to receive Jesus Christ. In place of that peace which religion is meant to give, these souls make their lives miserable, full of trouble and temptation. Scrupulous people distress themselves in many ways; for, really, they believe no one, and no counsel brings calm to their troubled souls. They keep returning to their sins and doubts, and the more they think of them the more they aggravate the trouble.   BL. HENRY SUSO

When they begin to hate sin and to amend themselves according to the laws of Holy Church, still there persists a fear that moves them to look at themselves and their sins committed in the past. And they take this fear for humility, but it is a reprehensible blindness and weakness.

BL. JULIAN OF NORWICH

After confessing his scruples returned, each time becoming more minute, so that he became quite upset, and although he knew that these scruples were doing him much harm and that it would be good to be rid of them, he could not shake them off. Sometimes he thought the cure would be for the confessor to tell him in the name of Jesus Christ never to mention anything of the past, and he wished that his confessor would so direct him, but he did not dare tell the confessor so.   IGNATIUS OF LOYOLA

Uneasy about your prayers? I forbid you to repeat even small parts of them.

Troubled over examination of conscience? Omit the daily examen.

You want to go to confession every day? No, once a week is enough. It's useless to make long confessions; just answer my questions and no more.

You're afraid you didn't have contrition for your sins? That regret indicates that you were sorry.

You went to confession to someone else last week after making your confession to me? I refuse you absolution today. Come around in a week.

You didn't go to communion as often as I told you to? I refuse to authorize you to receive at all this week; you're not worthy; when you learn to be more obedient, I'll tell you when you may go again.

You're troubled about your preparation for Communion? Don't make any. You should always be ready to communicate. Just see if you have any venial sins, make an act of contrition, kiss a crucifix, and receive without fear.

JOSEPH CAFASSO

When doubts and scruples of conscience assail you, you yourself are ignorant, unbalanced, and incapable of forming a judgement. Believe that . . . Possibly you will have to put up with the fears that torment you until you die. There is only one course of action—go ahead in blind obedience.

ALPHONSUS LIGUORI

Scruples . . . serve as purgatives—very active ones sometimes—to a soul which has just arisen from sin. They are useful to him for some time, and inspire him with fear and aversion as regards even the shadow of sin.

IGNATIUS OF LOYOLA

We may conclude that persons who suffer from scruples are the most favored by divine love, and the most certain of reaching Heaven when they bear this trial in patience and humility. Scrupulous souls die continually, they suffer a perpetual purgatory, and so they leave the earth to fly to Heaven purified and free from sins to expiate.   BL. HENRY SUSO

The path by which I travel isn't one of scrupulous fear; anything but that. I can always find some reason to be glad of my failures and make the best of them. And our Lord doesn't seem to mind; or why does he encourage me to follow that path? THÉRÈSE DE LISIEUX

Our infernal enemy observes with malignant attention what the stamp of our conscience is, whether it is delicate or relaxed. If delicate, he tries to render it more susceptible still; he endeavors to reduce it to the last degree of trouble and anguish, so as to stop its progress in the spiritual life.

IGNATIUS OF LOYOLA

Leave something for the angels. PHILIP NERI

God commands you to pray, but he forbids you to *worry*. JOHN VIANNEY

PENANCE

The concept of penances that go beyond the easily grasped idea of reparation for one's own sins is closely tied in with doctrines of the communion of saints and the mystical body of the Church. In this economy, the sacrifices of a few may ransom the many through voluntary penance. Holy women seem to have been especially eager to pay this price and to invite suffering to help balance the equilibrium between good and evil. "Were it not for this effort the universe might long ago have been in ruins."

Eternal God, accept the sacrifice of my life for the mystical body of thy holy Church.   CATHERINE OF SIENA

I want to suffer more, but I can't, for my confessor won't allow it. He forbids it, Lord, because I was taken sick in church this morning. But I am still strong enough, Jesus, and I want to help you and suffer more with you. You have spared neither blood nor life to save me, and I want to die for you.   GEMMA GALGANI

Suffering was now the magnet that drew me to itself. It had a charm that thrilled me, although I had never experienced it.   THÉRÈSE DE LISIEUX

While he was still at his studies in Barcelona, the desire returned of resuming his past penances, and he began by making a hole in the sole of his shoes, which widened little by little until by the time the cold of winter had arrived, nothing remained of the shoes but the uppers.

IGNATIUS OF LOYOLA

Inhabitants of Cortona, awake, and without wasting time, chase me from your region with stones, because I am the sinner who is guilty of excesses of all kinds.   MARGARET OF CORTONA

In order that I might make known my dissembling and my sins, it came into my mind to go throughout the cities and open places with meat and with fishes hanging about my neck and to cry: "This is that woman, full of evil and of dissembling, slave of all vices and iniquities, who did good deeds that she might obtain honor amongst men!" BL. ANGELA OF FOLIGNO

The penances done by some persons are as carefully ordered as their lives. They observe great discretion in their penances, lest they should injure their health. You need never fear that they will kill themselves; they are eminently reasonable folk! Their love is not yet ardent enough to overwhelm their reason. TERESA OF AVILA

Some burden themselves with indiscreet penances and with many other disorderly exercises of their own self-will, putting their confidence in such acts and believing they can thereby become saints. If they used half the same diligence in mortifying unruly appetites and passions they would advance more in one month than in many whole years with the other exercises. JOHN OF THE CROSS

Some are beguiled with overmuch abstinence from meat and drink and sleep. That is a temptation of the devil, to make them fall in the midst of their work, so that they bring it not to an end as they should have done, if they had known reason and kept discretion. RICHARD ROLLE

PRIDE

Pride is the primal sin and the most grave of all the sins, the spiritual rebellion that untuned the harmony of the universe. Pride is also the most glamorous, pervasive, fatal, and insidious of the sins. The saints have to be on guard as they develop, not to pride in their own virtues and the graces that they have received—lest all be turned to evil and they follow Lucifer in the fall from bright to dark.

If the thought which comes to you (or which you invite) is full of human conceit regarding your honor, your intelligence, your gifts of grace, your status, talents, or beauty, and if you willingly rest in it with delight, it is the sin of Pride. ANNONYMOUS, *The Cloud of Unknowing*

Knowing that the devil fell from Heaven through pride, for this cause the demons attack first those who have attained to a very great measure, seeking by means of pride and vainglory to turn them against one another. They know that in this way they can cut us off from God. ANTHONY OF EGYPT

When the spiritual and devout person feeleth himself to be greatly beloved of God and hath spiritual gifts and the works thereof, and doth openly speak of them, because he is too sure of himself and hath exceeded the right manner, God doth permit him to be in some way deceived in order that he may thereby learn better to know both God and himself.

BL. ANGELA OF FOLIGNO

Grace does not wish to be praised. And vice does not wish to be despised— that is, a man who has grace does not want to be praised and does not seek praise. And the man who has vices does not want to be despised or blamed— which comes from pride. BL. GILES OF ASSISI

My head was quite swollen with vanity, and my tainted heart was flattered at hearing all the praises and compliments I received.

ANTHONY MARY CLARET

An old man, very experienced in these matters, once spiritually admonished a proud brother who said in his blindness: "Forgive me, father, but I am not proud." "My son," said the wise old man, "What better proof of your pride could you have given than to claim that you were not proud?"

JOHN CLIMACUS

Do not think that your crosses are tremendous, that they are tests of your fidelity to God and tokens of God's extraordinary love for you. This thought has its source in spiritual pride. It is a snare quite subtle and beguiling but full of venom. You ought to acknowledge that you are so proud and sensitive that you magnify straws into rafters, scratches into deep wounds, rats into elephants.   LOUIS-MARIE GRIGNION DE MONTFORT

Pride makes us hate our equals because they are our equals; our inferiors from the fear that they may equal us; our superiors because they are above us. Envy, my children, follows pride; whoever is envious is proud.

JOHN VIANNEY

The more I see the difficulty of getting rid of this pest of pride, the more do I see the necessity of setting to work at it from childhood. For I find no other reason why this evil clings so to our hearts than because almost as soon as we are born it is sown in the tender minds of children by their nurses, it is cultivated by their teachers, and brought to its full growth by their parents.   THOMAS MORE

Men can heal the lustful. Angels can heal the malicious. Only God can heal the proud.   JOHN CLIMACUS

PHARISEES
AND HYPOCRITES

———————

The temptation to self-righteousness and hypocrisy is the temptation that attacks when others have been vanquished. Organized and triumphant religion is the greatest breeder of this evil. Both Christ and the saints condemn this vice with exceptional harshness. "This people honoreth me with their lips, but their hearts are far form me" (Mark 7:6). It is the one that assails the best educated, the cleverest, the leading citizens, the "most religious," "the holiest."

Jesus found his followers among the common people for the most part because the scribes and Pharisees were so sure that they had the answers that they despised Truth when it stood before them. Hence his repeated insistence on the necessity for childlike simplicity if one is to enter the Kingdom of Heaven. Anyone who must deal with the religious or civil hierarchy, as he and most of the saints have done, meets pharisaism in epidemic form. It is also a hazard on the path to sainthood, a particularly subtle and ugly threat to the development of the saints themselves.

———————

To pass judgement on another is to usurp shamelessly a prerogative of God, and to contemn is to ruin one's soul.   JOHN CLIMACUS

Again the fiend will deceive some people with another insidious plot. He will fire them with a zeal to maintain God's law by uprooting sin from the hearts of others. Never will he come right out and tempt them with something obviously evil. Instead, he incites them to assume the role of a zealous prelate supervising every aspect of the Christian life.

ANONYMOUS, *The Cloud of Unknowing*

Under the appearance of an austere life they have preserved the full strength of their passions. They have fed and made their own wills strong.

BL. HENRY SUSO

"God told me, God replied to me," they assert and yet most of the time they are talking to themselves. JOHN OF THE CROSS

It is of no use offering them advice, for they have been practicing virtue for so long that they think they are capable of teaching others and have ample justification for feeling as they do. TERESA OF AVILA

We are full of words but empty of actions, and therefore are cursed by the Lord, since he himself cursed the fig tree when he found no fruit but only leaves. It is useless for a man to flaunt his knowledge of the law if he undermines its teaching by his actions. ANTHONY OF PADUA

Nothing is more unworthy of a Christian, whose life should be an imitation of a God who is the soul of Honor and Truth itself, than to think one thing and say another. JOHN VIANNEY

I am greatly troubled and vexed in my spirit; for we wear the habit and have the name of saints, and boast of this before unbelievers. And I fear lest the word of Paul be fulfilled in us which says, "having the form of godliness but denying the power thereof." ANTHONY OF EGYPT

A man given to fasting thinks himself very devout if he fasts, although his heart may be filled with hatred. Much concerned with sobriety, he doesn't dare to wet his tongue with wine or even water but won't hesitate to drink deep of his neighbor's blood by detraction and calumny. FRANCIS DE SALES

My daughter, I see more Pharisees among the Christians than there were around Pilate. They renew My body's wounds so that, if that body were as large as the world, there would not be a spot the size of a point of a needle which would not be lacerated by their crimes. More Pharisees crucify Me today than at the time of My Passion. MARGARET OF CORTONA

It is not hypocritical if one's deeds fail to match one's words. Good gracious! Where should we be if it were? I should have to hold my tongue to avoid playing the hypocrite, since it would follow that were I to speak of perfection I should assume I was perfect. Certainly not, my very dear child. I no more believe I am perfect because I talk about perfection than I should believe myself Italian because I speak Italian, FRANCIS DE SALES

Let us look at our own shortcomings and leave other people's alone; for those who live carefully ordered lives are apt to be shocked at everything, and we might well learn very important lessons from the persons who shock us. TERESA OF AVILA

LUST AND
FORNICATION

John Climacus notes that people who commit other kinds of sin are said to have "slipped," but in fornication they "fall." All cultures are respectful and fearful of sexual desire because of the unique strength of this passion and because of the emotional and biological powers it unleashes, and all societies struggle to channel and control it. Our own society seems to be experimenting to determine the absolute minimum level of taboo and control we can retain and still survive.

The Church recognizes a distinction between sexual desire, which is natural and therefore morally neutral, and lust, which is, in effect, the idolization of sex, an inordinate desire for sexual or carnal pleasure. Chastity is the virtue opposed to lust and applies to all states of life, including marriage. The attitude of the Church toward sexuality is perhaps more dependent on the mores of society in general and subject to the *Zeitgeist* than are other major areas of doctrine.

Christ much preferred fornicators to Pharisees and told the chief priests and the elders that the harlots would go into the kingdom of God before them (Matthew 21:31).

Illicit desire for carnal indulgence or for the favor and flattery of others is called Lust. ANONYMOUS, *The Cloud of Unknowing*

Out of a forward will lust had sprung; and lust pampered had become custom; and custom indulged had become necessity. These were the links of the chain; this is the bondage in which I was bound.
AUGUSTINE OF HIPPO

For the sins of thy hands and arms, with which thou hast done much wickedness in embraces, touches, and other evil deeds, My hands were driven into the wood of the Cross by large nails and torn through bearing the weight of My body in Mine agony. BL. ANGELA OF FOLIGNO

Inordinate love of the flesh is cruelty, because under the appearance of pleasing the body we kill the soul.   BERNARD OF CLAIRVAUX

Our relentless enemy, the teacher of fornication, whispers that God is lenient and particularly merciful to this passion, since it is so very natural. Yet if we watch the wiles of the demons we will observe that after we have actually sinned they will affirm that God is a just and inexorable judge. They say one thing to lead us into sin, another thing to overwhelm us in despair.   JOHN CLIMACUS

Doubtless the state of virginity and continence is more perfect, but this does not prevent marriage from being holy, upright, and perfect in its degree, nor does it prevent those who live in marriage with true fear and love of God from being perfect, upright, and holy.   ANTHONY MARY CLARET

POSSESSIONS

The excessive or inordinate desire for and attachment to money and material things is the sin of avarice or covetousness, which leads to failures of charity, preoccupation with worldly cares to the neglect of things of the spirit, and sometimes to dishonesty, miserliness, or injustice. The story of the rich young man who was too much owned by his possessions so that he was not free to follow Jesus illustrates "how hard it is for them that trust in riches to enter into the kingdom of God" (Mark 10:17–25). It is a form of idolatry and is opposed by the attractive virtue of liberality.

If it is a thought of some material thing, that is of wealth or property or other earthly goods that people strive to possess and call their own, and if you dwell on it with desire, it is the sin of Covetousness.

ANONYMOUS, *The Cloud of Unknowing*

Avarice is an inordinate love of riches and the good things of this life. Jesus Christ, to cure us of it, was born in extreme poverty, deprived of all comforts. He chose a Mother who was poor. He willed to pass as the son of a humble workman.

JOHN VIANNEY

The love of worldly possessions is a sort of birdlime, which entangles the soul and prevents it flying to God.

AUGUSTINE OF HIPPO

Avarice is a worship of idols and is the offspring of unbelief. It makes excuses for infirmity and is the mouthpiece of old age. It is the prophet of hunger and the herald of drought. The miser sneers at the Gospel and is a deliberate transgressor. The man of charity spreads his money about him, but the man who claims to possess both charity and money is a self-deceived fool.

JOHN CLIMACUS

Riches are the instrument of all vices, because they render us capable of putting even our worst desires into execution.

AMBROSE

Every earthly possession is but a sort of garment for the body, and therefore he who hastens to contend with the devil should throw aside these
garments, lest he be borne down.   GREGORY I THE GREAT

Because of the sin of thy wealth, wherewith thou has done evil by acquir-
ing, wrongfully spending, and saving it, I have been poor, possessing neither palace nor house, nor hut, wherein I might be born or where I might
dwell during My lifetime; in death I should have had no sepulcher wherein
I might rest (but should have been left a prey unto dogs and birds) if one
had not been moved through compassion of My misery to receive Me into
his own sepulcher.   BL. ANGELA OF FOLINGO

Apothecaries have almost all kinds of poison for their use, as circumstances
may require, but they are not poisoned, because they keep their poisons
not in their bodies, but in their shops. In like manner you may possess
riches without being poisoned by them, provided you have them for use in
your house or in your purse, and not, by love, in your heart.
                                                              FRANCIS DE SALES

I shall never cease to give all I can to those in need until I find myself
reduced to such a state of poverty that there will scarcely remain to me five
feet of earth for my grave or a penny for my funeral.   CAJETAN

For those in the married state, the best example we can cite is that of St.
Joachim and St. Anne, who every year divided their income into three
equal parts. One was for the poor, the second for the temple and the divine
service, and the third for themselves.   IGNATIUS OF LOYOLA

Our Lord's given me the grace to care as little about gifts of the mind and
the heart as about worldly possessions. An idea occurs to me and I say
something which is well received by the other sisters—why shouldn't they
adopt it as their own? This idea doesn't belong to me, it belongs to the
Holy Spirit. To suppose it belongs to me would be to make the same mistake as the donkey carrying the relics, which imagined that all the reverence
shown to the Saints was meant for its own benefit!   THÉRÈSE DE LISIEUX

When you have got a psalter you will begin to want to have a breviary.
And when you have possessed yourself of a breviary you will sit on a high
chair like a great prelate and say to your brother: fetch me my breviary.
                                                           ANTHONY OF PADUA

Let others add coin to coin, fastening their claws on married ladies' purses
and hunting for deathbed legacies; let them be richer as monks than they
were as men of the world; let them possess wealth in the service of a poor

Christ such as they never had in the service of a rich devil; let the Church sigh over the opulence of men who in the world were beggars. Our virtuous Nepotian tramples gold under foot, books are the only things he desires.

JEROME

The superfluous riches which thou didst hoard and suffer to become rotten when thou shouldst have given them in alms to the poor, the superfluous garments which thou didst possess and preferred to see eaten by moths rather than clothing the poor, and the gold and silver which thou didst choose to see lie in idleness rather than spent on food for the poor, all these things, I say, will bear testimony against thee in the Day of Judgement.

ROBERT BELLARMINE

FOOD

Saints seem to be overwhelmingly vegetarian and very fond of fasting. It is not uncommon to read of saints subsisting for years on the Eucharist alone.

There have been few fat saints. Of these, the most famous is Thomas Aquinas. His size, combined with his absent-mindedness and his humility, caused him to be nicknamed "the Dumb Ox" during his early career. Later stories of his behavior at the endless series of testimonial and ceremonial banquets, which public celebrities must attend in all ages, make it clear that his girth was the result of his tendency to fall into deep abstraction in the course of these occasions, not noticing whether he was eating, talking to himself, or spilling gravy on the king of France.

Gluttony is hypocrisy of the stomach. Filled, it moans about scarcity; stuffed and crammed, it wails about hunger. Gluttony thinks up seasonings, creates sweet recipes. Stop up one urge and another bursts out.

JOHN CLIMACUS

There is another motion, when a man stuffs his body with food and drink and the heat of the blood from the abundance of nourishment rouses up warfare in the body because of our greed.   ANTHONY OF EGYPT

As time passed, the Soul lost more and more of its instinct for things divine. It, too, like the body, found satisfaction in the food of pigs and animals. All three were getting along splendidly together. As they all fared forth, happy and united, with no word of dissent, you can just imagine what became of higher reason. No one spoke of it; all their interest was focused on worldly things, and spiritual matters lost their appeal.

CATHERINE OF GENOA

Because of the sins of they mouth and throat, wherewith thou didst take delight in feasting and drunkeness and in the sweetness of delicate meats, My mouth hath been dry and empty, hungry and thirsty, it hath fasted and been made bitter with vinegar mingled with gall. BL. ANGELA FOLINGO

Elias the Thebite offers us an excellent example of frugality when he sat down beneath the juniper tree and the angel brought him food. "There was a hearth cake and a vessel of water." The Lord sent that sort of meal as the best sort for him. It seems then that we should travel light on our road toward truth.   CLEMENT OF ALEXANDRIA

Privies are the silent witnesses to the uncleanness of gluttony, for they are the depositories of the remains of the stomach's feasting.
                                                          CLEMENT OF ALEXANDRIA

Food ought to be a refection to the body and not a burden. BONAVENTURE

Irrational feeding darkens the soul and makes it unfit for spiritual experiences.   THOMAS AQUINAS

Take even bread with moderation, lest an overloaded stomach make you weary of prayer.   BERNARD OF CLAIRVAUX

I abstain from flesh lest I should cherish the vices of the flesh. A man becomes a beast by loving what beasts love.   BERNARD OF CLAIRVAUX

One ought to arise from a meal able to apply oneself to prayer and study.
                                                                           JEROME

The repletion of the stomach is the hotbed of lust.   JEROME

A wanton horse and an unchaste body should have their provender cut down.   HILARION

It is almost certain that excess in eating is the cause of almost all the diseases of the body, but its effects on the soul are even more disastrous.
                                                          ALPHONSUS LIGUORI

The success of your morning meditation will largely depend on what you have eaten the night before.   ALPHONSUS LIGUORI

Abstinence is the mother of health. A few ounces of privation is an excellent recipe for any ailment   BL. ANTHONY GRASSI

Lenten fasts make me feel better, stronger, and more active than ever.
                                                          CATHERINE OF GENOA

It is impossible to engage in spiritual conflict unless the appetite has first been subdued. GREGORY I THE GREAT

Continual moderation is better than fits of abstinence interspersed with occasional excesses. FRANCIS DE SALES

You should remember that frequently a demon can take up residence in your belly and keep a man from being satisfied even after having devoured the whole of Egypt and after having drunk all of the Nile. After we have eaten, this demon goes off and sends the spirit of fornication against us, saying: "Get him now! Go after him! His stomach is full and he won't put up much of a fight." JOHN CLIMACUS

Less care is required about bread than about other food, because it is less pleasing to the palate, and exposes us less to temptation. . . . Abstinence should be observed more particularly with regard to exquisite and rare meats, because they stimulate to concupiscence and provoke temptation.
IGNATIUS OF LOYOLA

Never did St. Dominic, even on his journeys, eat meat or any dish cooked with meat, and he made his friars do the same. BL. JORDAN OF SAXONY

It is useful for those who are weak in faith or unstable to abstain from certain meats, especially the palatable ones, because they have not enough faith in the protection of God. As says the Apostle, "One who believeth may eat all things, but he that is weak, let him eat herbs." NILUS SORSKY

We have bread, salt, butter and potatoes, and we are the happiest women in Ghent. JULIA BILLIART

He ate once daily, after sunset, but there were times when he received food every second and frequently even every fourth day. His food was bread and salt, and for drinking he took only water. There is no reason even to speak of meat and wine, when indeed such a thing was not to be found among the other zealous men. ATHANASIUS, *The Life of Anthony*

In general, give the body rather too much food than too little. PHILIP NERI

MORTIFICATION

Mortification is any conscious form of self-denial whose aim is to struggle against our evil inclinations and subject them to the rule of our will and the will of God. It is a Christian asceticism designed to master one's sinful tendencies and confirm one in the practice of virtue. When performed with a supernatural motive, mortification seeks by faith to grow in holiness by cooperating with the grace of God. Its value depends on motive; giving up a favorite food to lose weight is a different kind of act from giving up the same food out of love of God.

Mortification differs from penance insofar as it is a training of the will rather than a sacrifice made in atonement for sins committed by oneself or others.

As there should never be a truce between my soul and my body, I have decided not to spare the latter; let me tear it to pieces, mortify it, up to the very last moment of my life, when I shall at last see it separated from my soul. Do not think that it is as weak and mortified as it seems. It puts on a sham apathy in order to force me once more to taste the delights and pleasure to which I had accustomed it in the past.   MARGARET OF CORTONA

My mind was faced with choosing between my pleasure and God's, and since my mind saw the glaring inequality between the two, even in the slightest matter, I would be forced to choose what then seemed more pleasing to God.   ANTHONY MARY CLARET

All for God and nothing for self.   MARY MAGDELENE DEI PAZZI

The reason for which it is necessary for the soul, in order to attain Divine Union with God, to pass through this dark night of mortification of the desires and denial of pleasures in all things, is because all the affections which it has for creatures are pure darkness in the eyes of God, and, when the soul is clothed in these affections, it has no capacity for being enlightened and possessed by the pure and simple light of God if it cast them not first from it; for the light cannot agree with darkness.

JOHN OF THE CROSS

The body is brought under the authority of the mind, being taught by the spirit, as St. Paul says: "I keep under my body, and bring it into subjection." For the mind purifies it from food and from drink and from sleep, and in a word from all its motions, until through its own purity it frees the body even from the natural emission of seed.   ANTHONY OF EGYPT

When he was thinking of the things of this world he was filled with delight, but when afterwards he dismissed them from weariness, he was dry and dissatisfied. And when he thought of going barefoot to Jerusalem and of eating nothing but herbs and performing other rigors he saw that the saints had performed, he was consoled, not only when he entertained these thoughts but even after dismissing them he remained cheerful and satisfied.

IGNATIUS OF LOYOLA

I send you this hairshirt to use when you find it difficult to recollect yourself at times of prayer, or when you are anxious to do something for the Lord. . . . It can be worn on any part of the body and put on in any way so long as it feels uncomfortable. . . . It makes me laugh to think how you send me sweets and presents and money, and I send you hairshirts.

TERESA OF AVILA, to her brother

I did dispose and determine that, even though I should be forced to die of hunger, cold, and shame, because such a thing was pleasing or might be pleasing to God I would by no means leave from my purpose, even though I were certain that these aforesaid evils should befall me, choosing to die willingly for the love of God rather than to fall short of my intention.

BL. ANGELA OF FOLIGNO

Do you not know that fasting can master concupiscence, lift up the soul, confirm it in the paths of virtue, and prepare a fine reward for the Christian?

HEDWIG OF SILESIA

Spurred on by his heavenly vision of the joys of eternal bliss, Cuthbert was ready to suffer hunger and thirst in this life in order to enjoy the banquets of the next.   BEDE THE VENERABLE

The mortified man is able to suck honey from the rock and oil from the rugged stones.   BERNARD OF CLAIRVAUX

Fight the good fight, my daughters, against our ancient foe, fight him insistently with fasting, because no one will win the crown of victory without engaging in the contest in the proper way.   DOMINIC

The spiritual combat in which we kill our passions to put on the new man is the most difficult of all the arts.   NILUS THE ABBOT

The more you hate and ill-treat your flesh, the greater will be your reward in the next life.   BENEDICT JOSEPH LABRE

The best way not to find the bed too cold is to go to bed colder than the bed is.   CHARLES BORROMEO

As for austerities, I am afraid I am not given to hair-shirts, sleeping on the ground, a bread and water diet, etc., for as I am now hastening on toward my sixtieth year and my health is all but broken, I doubt whether I could support such hardships for long. Still, if ever a spiritual and prudent man should recommend them, I think, unless my self-love is playing me a trick, I would be quite ready to take them up.   ROBERT BELLARMINE

I don't mean to suggest that I went in for penitential practices of any kind. That's a thing, I'm afraid, I've never done; I've heard so much about saintly people who took on the most rigorous mortifications from their childhood upwards, but I'd never tried to imitate them. The idea never had any attractions for me. I expect that comes from cowardice on my part.

THÉRÈSE DE LISIEUX

They who pay a moderate attention to the mortification of their bodies and direct their main attention to mortify the will and understanding, even in matters of the slightest moment, are more to be esteemed than those who give themselves exclusively to bodily penances.   PHILIP NERI

Certain virtues are greatly esteemed and always preferred by the general run of men because they are close at hand, easily noticed, and, in effect, material. Thus many people prefer bodily to spiritual alms, hair shirts, fasting, going barefoot, using the discipline, and physical mortifications to meekness, patience, modesty, and other mortifications of the heart although the latter are really higher virtues.   FRANCIS DE SALES

There are people who are very mortified exteriorly but they have not renounced themselves in God. . . . These people do not wish to conform their lives to that of Jesus Christ, which was gentleness and humility itself. They judge and blame their neighbor all too easily. They despise and condemn all who do not live as they do; and if one wishes to know them as they really are one has only to wound them in their self love and the good

opinion they have of themselves. Then they are found to be full of pride
and disquietude.   BL. HENRY SUSO

The immoderate long fasts of many displease me, for I have learned by
experience that the Ass that is too much fatigued on the road seeks rest at
any cost. In a long journey, strength must be supported.   JEROME

The soul should treat the body as its child, correcting without hurting it.
                                    FRANCIS DE SALES

The practices of mortification should be moderated by prudence and the
advice of a wise director because it often happens that the devil urges a soul
to excessive penances to tire her and render her unfit for the service of God
and the fulfillment of her duties.   BL. ANNA MARIA TAIGI

It is your self-love that leads you to hate and avoid anything that might
cause pain or mortification to your spirit or your flesh and makes you love
and seek out everything that may give them pleasure and contentment.
                                    JEAN EUDES

The perfection of a Christian consists in mortifying himself for the love of
Christ. Where there is no great mortification, there is no great sanctity.
                                    PHILIP NERI

SCRIPTURES

Holy Scripture is the name given to the collection of the sacred books of the Jews and Christians. It is a source of signs, prophetic guidance, inspiration, profound understanding, and sound precept. The saints insist that the light of grace is necessary to profit from Scripture; knowledge without grace is dead, and the letter without the spirit kills.

If you observe anything evil within yourself, correct it; if something good, preserve it; if something beautiful, foster it; if something sound, maintain it; if sickly, heal it; Read unwearingly the precepts of the Lord and, sufficiently instructed by them, you will know what to avoid and what to pursue.   BERNARD OF CLAIRVAUX

Every creature is by its nature a kind of effigy and likeness of the eternal Wisdom, but especially one which in the Scripture has been elevated by the spirit of prophecy to prefigure spiritual things.   BONAVENTURE

Just as at the sea those who are carried away from the direction of the harbor bring themselves back on course by a clear sign, on seeing a tall beacon light or some mountain peak coming into view, so Scripture may guide those adrift on the sea of life back into the harbor of the divine will.
                                                                GREGORY OF NYSSA

What moved and stimulated me most was reading the Holy Bible, to which I have always been very strongly attracted. There were passages that impressed me so deeply that I seemed to hear a voice telling me the message I was reading.   ANTHONY MARY CLARET

The collection of psalms found in Scripture, composed as it was under divine inspiration, has, from the very beginnings of the Church, shown a wonderful power of fostering devotion among Christians.   PIUS X

The psalms seem to me to be like a mirror, in which one can see himself and the stirrings of his own heart; he can recite them against the background of his own emotions. ATHANASIUS

How I wept when I heard your hymns and canticles, being deeply moved by the sweet singing of your Church. Those voices flowed into my ears, truth filtered into my heart, and from my heart surged waves of devotion.

AUGUSTINE OF HIPPO

Everything that we read in the sacred Books shines and glitters even in the outer shell; but the marrow is sweeter. He who desires to eat the kernel must first break open the shell. JEROME

Every light that comes from Holy Scripture comes from the light of grace. This is why foolish, proud and learned people are blind even in the light, because the light is clouded by their own pride and selfish love. They read the Scripture literally, not with understanding. They have let go of the light by which the Scripture was formed and proclaimed.

CATHERINE OF SIENA

By nothing that we can think or say can God be exalted. The Holy Scriptures are so far above us that no man—be he the wisest in all the world and possessing all the knowledge it is possible to have in this life—can fully and perfectly know and understand them. There is none whose intelligence would not be always overcome by them. BL. ANGELA OF FOLINGO

I am pleased for my friars to study the Scriptures as long as they do not neglect application to prayer, after the example of Christ, of whom we read that he prayed more than he read. FRANCIS OF ASSISI

KNOWLEDGE
AND REASON

We can know God by the light of natural human reason because "ever since God created the world his everlasting power and deity—however invisible—have been there for the mind to see in the things he has made" (Romans 1:20). As St. Thomas Aquinas tells us, if the power of reason is rightly applied, it must lead toward divine truth, even in the case of those geniuses who may have been deprived by historical accident of more direct and accurate ways of knowing.

The knowledge that comes from God's revelation and the inspiration of the Holy Spirit, however, are of a higher order; "divine treasures which totally transcend the understanding of the human mind." Because of the gifts of revelation, ordinary people who are not great philosophers or geniuses can possess true and accurate knowledge of divine things.

Reason is a useful tool, like many others, but it cannot be expected to compare with revelation as a way to understanding truth. Human intelligence and knowledge are limited and often clouded by "the darkness of sin," particularly that of pride.

Poor human reason when it trusts in itself substitutes the strangest absurdities for the highest divine concepts. JOHN CHRYSOSTOM

Reason and human science often lead you into error because they are too weak and limited to penetrate to the knowledge of the things of God, which are infinite and incomprehensible. Human intelligence and knowledge also deceive you, because they are too full of the darkness and obscurity of sin to attain to a genuine knowledge even of things outside of God.

JEAN EUDES

Learning unsupported by grace may get into our ears; it never reaches the heart. But when God's grace touches our innermost minds to bring under-

standing, his word which has been received by the ear sinks deep into the
hears.   ISIDORE OF SEVILLE

We Christians, then, do not possess the mystery through wisdom based on
Greek reasoning, but in the power given to us by God, through Jesus Christ.
As evidence that this is true, you can see that although we are unlettered,
we do believe in God and know through his works that his providence cares
for all things. We rely on Christ for the truth of our faith, while you rely
on sophistry and clever words.   ANTHONY OF EGYPT, *to the Greeks*

Faith is the proof of what cannot be seen. What is seen gives knowledge,
not faith. When Thomas saw and touched, why was he told, "You have
believed because you have seen me?" Because what he saw and what he
believed were different things. God cannot be seen by mortal man.
                                                  GREGORY I THE GREAT

No matter how enlightened one may be through natural and acquired
knowledge, he cannot enter into himself to delight in the Lord unless Christ
be his mediator.   BONAVENTURE

The human intellect has a greater desire, and love, and pleasure, in knowing divine matters than it has in the perfect knowledge of the lowest things,
even though it can grasp but little concerning divine things. So, the ultimate end of man is to understand God, in some fashion.   THOMAS AQUINAS

Christ is the first born of God, and he is the Word of whom all mankind
have a share, and those who lived according to reason are Christians even
though they are classed as atheists. For example, among Greeks, Socrates
and Heraclitus; among non-Greeks Abraham, Ananias, Azarias, and Mis-
ael, and Elias, and many others.   JUSTIN MARTYR

You cannot imagine how foolish people are. They have no sense of discernment, having lost it by hoping in themselves and putting their trust in their
own knowledge. O stupid people, do you not see that you are not the
source of your own knowledge? It is my goodness, providing for your needs,
that has given it to you.   CATHERINE OF SIENA

Truly barren is profane education which is always in labor but never gives
birth. For what fruit worthy of such pangs does philosophy show for being
so long in labor? Do not all who are full of wind and never come to term
miscarry before they come into the light of the knowledge of God? They
could as well become men if they were not altogether hidden in the womb
of barren wisdom.   GREGORY OF NYSSA

If knowledge can cause most people to become vain, perhaps ignorance and lack of learning can make them humble. Yet now and then you do find men who pride themselves on their ignorance.   JOHN CLIMACUS

Who is there in this illustrious home of learning who does not think daily as he goes to the schools of law, medicine, philosophy, or theology, how best he may progress in his particular subject and at last win his doctor's degree? Well, the school of Christ is the school of charity. In the last day, when the great general examination takes place, there will be no question at all on the text of Aristotle, the aphorisms of Hippocrates, or the paragraphs of Justinian. Charity will be the whole syllabus.

ROBERT BELLARMINE

This year I read the works of St. Teresa of Avila once more, and the Lord blessed this reading with great gifts of knowledge. How good the Lord is! Since he foreknew the great trials I would have to undergo, he forearmed me with great insights and spiritual help.   ANTHONY MARY CLARET

The end of my labors has come. All that I have written appears to me as so much straw after the things that have been revealed to me.

THOMAS AQUINAS

SIMPLICITY

A most attractive virtue, simplicity opposes hypocrisy with an artless, unpretentious sincerity that has no hint of guile. There is a hint of vulnerability, a feeling that the simple are likely to be the prey of the clever. "I fear, lest by any means, as the serpent beguiled Eve through his subtilty, so your minds should be corrupted from the simplicity that is in Christ" (2 Corinthians 11:3). Christ seems to be defining an ideal degree of simplicity when he tells the apostles that he sends them forth as "sheep in the midst of wolves: be ye therefore wise as serpents, and harmless as doves" (Matthew 10:16).

As a supernatural virtue, simplicity single-mindedly wishes to do the will of God, without consideration of any advantage to self. The high value placed on simplicity comes ultimately from Christ's insistence that only those who are like children can enter the Kingdom.

God in his nature is most simple and cannot admit of any duplicity. If we then would be conformable to Him, we should try to become by virtue what He is by nature. We should be simple in our affections, intentions, actions, and words; we should do what we find to do without artifice or guile, making our exterior conformable to our interior. We should have no other object but God in our actions and seek to please Him alone in all things.   VINCENT DE PAUL

Christ appeared not as a philosopher or a doctor of many words, or as one who disputeth noisily, nor yet as a scribe renowned for wisdom and learning; but in the utmost simplicity did He talk with men, showing unto them the way of truth in His life, His virtues, and His miracles.

BL. ANGELA OF FOLIGNO

When our Lord sees pure souls coming eagerly to visit him in the Blessed Sacrament, he smiles on them. They come with that simplicity that pleases him so much. JOHN VIANNEY

This is the highest point of philosophy, to be simple and wise; this is the angelic life. For the soul of a little child is free from all diseases of the mind; he retains no memory of wrongs but approaches those who have inflicted them in friendly fashion, as if nothing had happened. And although he may be chastened by his mother with blows, he ever seeks her out, looking up to her above all others. JOHN CHRYSOSTOM

Even amongst the poor, while a child is still small, he is given what he needs, but once he is grown up, his father will no longer feed him but tells him to seek work and support himself. Well, it was to avoid hearing this that I wished never to grow up, for I feel incapable of earning my livelihood, which is Life Eternal. THÉRÈSE DE LISIEUX

Be sober and hard-working men; avoid all vanity in dress which will exclude you from Heaven; try to keep to the simplicity of manners of our fathers. NICHOLAS OF FLÜE, *to the Swiss*

I am a Lombard and I detest dissimulation. What I have in my heart I have also on my tongue. PAUL OF THE CROSS

It is a kind of candid and ingenuous confession to praise in others that which is wanting in ourselves. JEROME

Craftiness is the accumulation of artifices, intrigues, deceits, and appearances to mislead the minds of those with whom we converse. This is quite the reverse of simplicity, which requires that the outside should correspond with what is within. FRANCIS DE SALES

A humble person, if his opinion is asked, gives it in all simplicity and then leaves others to give theirs. Whether they are right or wrong, he says no more. JOHN VIANNEY

He willeth then that we use the condition of a child: for when it is diseased, or adread, it runneth hastily to the mother for help, with all its might. So willeth he that we do as a meek child saying thus: "My kind Mother, my Gracious Mother, my dearworthy Mother, have mercy on me: I have made myself foul." BL. JULIAN OF NORWICH

My political views are those of the Lord's prayer. JOHN BOSCO

HOLY FOLLY

"If any man among you seemeth to be wise in this world, let him become a fool, that he may be wise. For the wisdom of this world is foolishness with God. For it is written, He taketh the wise in their own craftiness. And again, the Lord knoweth the thoughts of the wise, that they are vain" (1 Corinthians 3:18–20). The true follower of Christ is a "fool" who rejects worldly values to embrace humiliation, suffering, poverty, and a martyr's death. The rich, the respectable, and the learned will enter the Kingdom, if at all, behind the harlots, the publicans, the beggars, the thieves, the sinners. "God hath chosen the foolish things of this world to confound the wise; and God hath chosen the weak things of the world to confound the things which are mighty" (1 Corinthians 1:27).

In addition to a reversal of worldly values, Holy Folly includes the element of unmeasured, reckless love of God and the willingness to go to wild extremes for the sake of this love. To be a Holy Fool is a special calling within the Christian tradition. Most of the great saints have at least a touch of this divine madness.

Out of gratitude and love for Him, we should desire to be reckoned fools and glory in wearing His livery.   IGNATIUS OF LOYOLA

One day perhaps you will see me with the public executioner behind me, whipping me, and you will say, "Good heavens, surely it's that Father Philip we used to think such a good little man."   PHILIP NERI

Our Lord might have used all the subtlety of knowledge and of argument, and, had he desired, He might have shown forth his wit and obtained glory; but with such simplicity did he declare the truth that he was esteemed of almost all people to be not only simple and foolish, but even ignorant and vain.   BL. ANGELA OF FOLIGNO

O my God, how can you reproach me? Is it not you who have taught me these follies? Have you not given me the first example of folly by imprisoning yourself for me?   GERARD MAJELLA

The great saints, in their eagle strength, have gone close to the verge of folly in the wonderful things they did for you; I am too poor a creature to do anything wonderful, but I must be allowed the folly of hoping that your love will accept me as its victim.   THÉRÈSE DE LISIEUX

To be a saint, one must be beside oneself. One must lose one's head.
                                                              JEAN VIANNEY

I pretended to be a fool, though without neglecting anything of my service of everyone. When those pitiless fathers saw that I willingly served in the same status, they loaded all the heavy work of the monastery onto me. I spent thirteen years this way, and then in a dream I saw those who had appeared to me before, and they gave me a receipt to mark full payment of my debt.   JOHN CLIMACUS, *of a saintly monk*

What else do worldlings think we are doing but playing about when we flee what they most desire on earth, and what they flee, we desire? We are like jesters and tumblers who, with heads down and feet in the air draw all eyes to themselves. . . . Ours is a joyous game, decent, grave, and admirable, delighting the gaze of those who watch from heaven. This chaste and religious game he plays who says, "We are made a spectacle to angels and to men."   BERNARD OF CLAIRVAUX

Every time I do not behave like a donkey, it is the worse for me. How does a donkey behave? If it is slandered, it keeps silent; if it is not fed, it keeps silent; if it is forgotten, it keeps silent; it never complains, however much it is beaten or ill-used, because it has a donkey's patience. That is how the servant of God must behave. I stand before you, Lord, like a donkey.
                                                              PETER CLAVER

The virtue of innocence is held as foolishness by the wise of this world. Anything that is done out of innocence, they doubtless consider to be stupidity, and whatever truth approves of, in practice is called folly by men of worldly wisdom.   GREGORY I THE GREAT

ILLUMINATION

Light and darkness are the most compelling and universal images of good and evil. Christ made full use of this imagery and so do the saints. Light is also very often used to symbolize understanding and revelation, which are at all times the gifts of the Holy Spirit, and make the fruits of human reason "appear as straw."

Since he is the Sun of Justice, he fittingly calls his disciples the light of the world. The reason for this is that through them, as through shining rays, he has poured out the light of the knowledge of himself upon the entire world. For by manifesting the light of truth, they have dispelled the darkness of error from the hearts of men.   CHROMATIUS OF AQUILEIA

Lord, shed upon our darkened souls the brilliant light of your wisdom so that we may be enlightened and serve you with renewed purity. Sunrise marks the hour for men to begin their toil, but in our souls, Lord, prepare a dwelling for the day that will never end. Through our unremitting zeal for you, Lord, set upon us the sign of your day that is not measured by the sun.   EPHRAEM OF NISIBIS

Just as God's creature, the sun, is one and the same the world over, so also does the Church's preaching shine everywhere to enlighten all men who want to come to a knowledge of truth.   IRENAEUS

Born of a virgin, he came forth from the womb as the light of the whole world in order to shine on all men. His light is received by those who long for the splendor of perpetual light that night can never destroy. The sun of our daily experience is succeeded by the darkness of night, but the sun of holiness never sets, because wisdom cannot give place to evil.   AMBROSE

Our lighted candles are a sign of the divine splendor of the one who comes to expel the dark shadows of evil and to make the whole universe radiant with the brilliance of his eternal light. Our candles also show how bright our souls should be when we go to meet Christ.

SOPHRONIUS, *on the Feast of the Presentation of the Lord*

You are a fire that takes away the coldness, illuminates the mind with its light, and causes me to know your truth. And I know that you are beauty and wisdom itself. The food of angels, you gave yourself to man in the fire of your love.   CATHERINE OF SIENA

Learn to fix the eye of faith on the divine word of the Holy Scriptures as on a light shining in a dark place until the day dawn and the day-star arise in our hearts. For the ineffable source from which this lamp borrows its light is the Light that shines in darkness but the darkness does not comprehend it. To see it, our hearts must be purified by faith.

AUGUSTINE OF HIPPO

Most high, glorious God, enlighten the darkness of my heart and give me, Lord, a correct faith, a certain hope, a perfect charity, sense, and knowledge, so that I may carry out your holy and true command.

FRANCIS OF ASSISI

As he sat, the eyes of his understanding began to open. He beheld no vision, but he saw and understood many things, spiritual as well as those concerning faith and learning. This took place with so great an illumination that these things appeared to be something altogether new.

IGNATIUS OF LOYOLA

"Hear, O Israel, the Lord your God is one." If you see this in the pure simplicity of your mind, you will somehow be bathed in the brilliance of eternal light.   BONAVENTURE

Those who are led by the Holy Spirit have true ideas; that is why so many ignorant people are wiser than the learned. The Holy Spirit is light and strength.   JOHN VIANNEY

When the soul is illumined by the presence of God and doth repose in God's bosom and God in it, then is it exalted above itself and heareth and rejoiceth and doth rest in that divine goodness, concerning which none can report because it is above all intelligence and all manner of speech and above all words. But herein doth the soul swim in joyfulness and in knowledge, and thus enlightened, it comprehendeth the meaning of all the difficult and obscure sayings of Christ.   BL. ANGELA OF FOLIGNO

DESOLATION

———

Desolation is the subjective feeling that one is separated from God. It has many stages, ranging from the "dryness" which is simply the absence of any sensible consolation or active sense of God's presence or approval, through various levels of melancholy, feelings of abandonment, loss of faith and purpose, and experience of the temptation to despair. Most saints are as liable as we are to experience desolation, but they are considerably more likely to recognize it for a trivial distraction or a temptation and to bear it better. If it is seen merely as one trial among many, it loses its dangers, if not its sting. To hear the most staunch and trusting saints describe their pain, loneliness, and doubt is both touching and inspiring.

———

The perfection of Christian detachment does not consist only in being detached from the world and from oneself. It obliges the soul to be, in a certain sense, detached even from God. "It was expedient for Him to depart from them."   JEAN EUDES

O God, you seek those who hide from you, and hide from those who seek you!   AUGUSTINE OF HIPPO

Without you, my sweet Savior, I remain in darkness and grief. Without you, most gentle Lamb, I remain in worry and fear. Without you, son of Almighty God, I remain in confusion and shame. For without you I am deprived of all that is good. Without you I am blind in the dark because you are Jesus, the true Light of the World. Without you I am lost and damned, because you are the God of Gods and the giver of graces.
                                                        BL. JOHN OF ALVERNA

I get tired of the darkness all around me and try to refresh my jaded spirits with the thoughts of that bright country where my hopes lie; and what happens? It is worse torment than ever; the darkness itself seems to borrow, from the sinners who live in it, the gift of speech. I hear its mocking accents; "It's all a dream, this talk of a heavenly country, . . . Death will make nonsense of your hopes; it will only mean a night darker than ever, the night of mere non-existence." THÉRÈSE DE LISIEUX

I seek Him and He won't be found; He turns a deaf ear to my sighs and moaning. "Tell me who you are and what you want," I say to Him. "Make yourself known and then let me die." I am uncontrolled almost to the extent of being rude to Him. I end up by calling Him cruel, but immediately afterward beg His pardon. Some of the things I say are prompted not by anger but by so much love. GEMMA GALGANI

When I was in that darkness of spirit methought I would have chosen rather to be roasted than to endure such pains. Wherefore did I cry aloud and call upon death, desiring that it should come in any form whatsoever if only God would permit me to die. And unto God did I say, "Lord, if Thou wilt send me into hell, I pray Thee tarry not, but do it instantly, and since Thou has abandoned me, make an end of it now and plunge me into the depths." BL. ANGELA OF FOLIGNA

I shall and shall not be with you. I shall clothe you in My Grace, but you will think yourself deprived of it, because while dwelling within you I shall be able to go unrecognized. I am concealing Myself from you so that you may discover by yourself what you are without Me.

MARGARET OF CORTONA

The cause of this dryness is that God is transferring to the spirit the goods and energies of the senses, which, being now unable to assimilate them, become dry, parched up, and empty, for the sensual nature of man is helpless in those things which belong to the spirit simply. Thus, the spirit having tasted, the flesh shrinks and fails. JOHN OF THE CROSS

Often it is the Lord's will that we should be persecuted and afflicted by evil thoughts, which we cannot cast out, and also by aridities; and sometimes He even allows these reptiles to bite us, so that we may learn better how to be on our guard in the future. TERESA OF AVILA

You can advance farther in grace in one hour during this time of affliction than in many days during a time of consolation. JEAN EUDES

Perfection does not consist in consolation, but rather in the submission of our wills to God, above all in desolation. Let us remember that the obedience of our Lord was perfect when his mouth and tongue burned with the fever of his wounds, and this was only increased and his thirst became greater when they gave him gall and vinegar to drink. BL. HENRY SUSO

Whenever the feeling of grace is withdrawn, pride is the cause. Not necessarily because one has actually yielded to pride, but because if this grace were not withdrawn from time to time pride would surely take root. God in his mercy protects the contemplative in this way, though some foolish neophytes will think he has turned enemy to them.

ANONYMOUS, *The Cloud of Unknowing*

If we are to rise above this depression, dejection and despondency of soul, and turn it to use in God's service, we must face it, accept it, and realize the worth of holy self-abasement. In this way you will transmute the lead of your heaviness into gold, a gold far purer than any of your gayest, most light-hearted sallies. Well, then be patient with yourself. See to it that your higher self puts up with your lower.

FRANCIS DE SALES

In time of desolation one should never make a change, but stand firm and consistent in the resolutions and decisions that guided him the day before the desolation, or to the decision which he observed in the preceding consolation. For just as the good spirit guides and rejoices us in consolation, so in desolation it is the evil spirit which guides and counsels.

IGNATIUS OF LOYOLA

When it comes to his closest friends—his Mother for example—he tests their faith by keeping them waiting for the miracle. He allows Lazarus to die when Martha and Mary have sent warning that he is sick, At Cana of Galilee when our Lady puts the difficulties of their host before him, he tells her that his time has not come yet. Why shouldn't our Lord treat me the same way, by keeping me waiting first, then satisfying all the dearest wishes of my heart?

THÉRÈSE DE LISIEUX

You desire that it should always be spring in your soul, but that cannot be. We must endure vicissitudes of weather interiorly as well as exteriorly. It is only in heaven that we shall find the perpetual ripening of summer, the perpetual fruition of autumn. There we shall have no winter, but here winter is required for the exercise of abnegation and of a thousand little virtues which are practiced in times of sterility.

FRANCIS DE SALES

Where now is the joy of your presence which I seek above everything?

MARGARET OF CORTONA

CONSOLATION

Consolation is the name given to a variety of pleasurable experiences involving a vivid sense of God's reality, approval, and immanence. Because the experience of consolation is both joyful and elusive, many saints face the subject with ambivalent feelings. The pursuit of consolation can lead to a form of "spiritual gluttony," and the sensations of spiritual sweetness and delight can become an end in themselves, blocking or diverting the development of the soul's relationship to God. Temptations sometimes disguise themselves as consolation to distract the unwary from profitable forms of prayer. In some cases, however, experience of the sweetness of consolation leads straight to holiness. God knows best who needs sensible consolations and when they need them. It seems appropriate to read of martyrs being "filled with a supernatural sweetness of spirit" when facing torture.

In anticipation of this eternal glory, God will sometimes inflame the senses of his devout friends with unspeakable delight and consolation even here in this life. And not just once or twice, but perhaps very often as he judges best. This delight, however, does not originate outside the person, entering through the windows of the faculties, but wells up from an excess of joy and true devotion of spirit.   ANONYMOUS, *The Cloud of Unknowing*

God poureth into the soul an exceeding great sweetness, in a measure so abundant that it can ask nothing more—yea, verily, it would be in Paradise if this should endure, its joy being so great that it filleth the whole body; and all injury that the soul suffereth, whether by words or deeds, is esteemed as naught and is turned into sweetness.   BL. ANGELA OF FOLIGNO

Here are great dishes of food and drink, of which no one knows save he who tastes them: but full satisfaction in fruition is the dish which is lacking there, and therefore this hunger is ever renewed. Yet, in the touch, rivers of honey, full of all delights, flow forth. The spirit tastes these riches in all the ways which it can conceive and comprehend, but all this is in a creaturely way and below God, and hence there remains an eternal hunger and impatience.   BL. JAN VAN RUYSBROECK

When all the joys of heaven come flooding into a human heart how difficult it is for that heart, still in exile, to stand the strain of the impact without finding relief in tears.   THÉRÈSE DE LISIEUX

The first day that I was tortured upon the rack, before I came to the torture chamber, giving myself up to prayer, I was filled with a supernatural sweetness of spirit, and even while I was calling upon the Most Holy Name of Jesus and upon the Blessed Virgin Mary (for I was saying my rosary) my mind was cheerfully disposed, well comforted, and readily inclined and prepared to suffer and endure these torments.   ALEXANDER BRIANT

I remember not ever to have tasted such interior delights, and these consolations of the soul are so pure, so exquisite, and so constant, that they take from me all sense of my corporal suffering.   FRANCIS XAVIER

And on another night—I do not know, God knows whether it was in me or beside me—someone was speaking in the most elegant language which I listened to but could not understand, except that at the end of the speech he spoke these words, "He who gave his life for you, he it is who speaks in you," and at that I woke up full of joy.   PATRICK OF IRELAND

O good Jesus, who shall grant me to feel Thee, who now neither may be felt nor seen? Shed Thyself into the entrails of my soul! Come into my heart and fill it with thy most excellent sweetness. Inebriate my mind with the hot wine of Thy sweet love, that, forgetting all evils and all scornful visions and imaginations, and having Thee alone I may be glad and rejoice in Jesus my God.   BL. RICHARD ROLLE

Ignorant would he be if he went after God in search of this sweetness, and rejoiced and rested in it; for in this case he would not be seeking God with his will grounded in the emptiness of faith and charity, but spiritual sweetness and pleasure, which is of creature, following his taste and desire; and thus he would not love God purely, above all things.   JOHN OF THE CROSS

Some people are so spiritually fragile and delicate that unless they were always strengthened with a little sensible consolation, they might be unable to endure the various temptations and sufferings that afflict them as they struggle in this life against their enemies from within and without.
                    ANONYMOUS, *The Cloud of Unknowing*

For often, as you have read, it is to the weakest that His Divine Majesty gives favors, which I believe they would not exchange for all the fortitude given to those who go forward in aridity. We are fonder of spiritual sweetness than of crosses.   TERESA OF AVILA

You shed a few crocodile tears in the course of your reading and thrilled with an emotion strange to you—well, there is nothing remarkable about that. Christ was merely enticing you with this sweet invitation to bear a little of his cross. As a rule, people who aim at a spiritual life begin with the sweet and afterwards pass on to the bitter.   PHILIP NERI

There are certain souls who are always looking for consolation in prayer; this is a delusion of the devil who simply wishes to bring about their destruction.   ALPHONSUS LIGUORI

Perfection does not consist in consolation, but rather in the submission of the will to God alone; submission above all in things that are hard and bitter.   BL. HENRY SUSO

When one has no consolations, one serves God for himself alone, but when one has them one is liable to serve him out of love for self.   JOHN VIANNEY

Some people experience a measure of consolation almost always while others only rarely. But God in his great wisdom determines what is best for each one.   ANONYMOUS, *The Cloud of Unknowing*

To these perfect souls trouble is a pleasure, and pleasure and every consolation that the world would offer them are a toil—and not only the consolation that the servants of the world, by my dispensation, give them in compassion for their bodily needs, but also even the mental consolation which they receive from Me. Even this they despise through humility and self hatred. They do not despise consolation itself, which is my gift and grace, but only the pleasure which the soul's appetite finds therein.
                    CATHERINE OF SIENA

As this heavenly water begins to flow from this source of which I am speaking—that is, from our very depths—it proceeds to spread within us and cause an interior dilation and produce ineffable blessings, so that the soul itself cannot understand all that it receives there.   TERESA OF AVILA

After death, true life; after desolation, true consolation; a life which delivers our souls from death and a consolation which restrains our eyes from tears.
                    AUGUSTINE OF HIPPO

JOY

Joy is an integral part of our relationship to God, the source of all good and all happiness, and is the fruit of his revelation. "These things have I spoken unto you, that my joy might remain in you, and that your joy might be full" (John 15:11). The closer we are to the Kingdom of God, the more intense our sharing in the rejoicing that is an essential action of the divine nature. Saints experience more joy on earth than others do because they are closer to God and more open to the gifts of the Holy Spirit. A mark of great holiness is the ability to find happiness in the most unpromising circumstances, to rejoice in deprivation, pain, and sacrifice. Even saints who make penance and suffering a major part of their career must do so joyfully or lose the game.

No one can live without delight and that is why a man deprived of spiritual joy goes over to carnal pleasures. THOMAS AQUINAS

I return from Perugia and arrive here in the dead of night; and it is winter time, muddy and so cold that icicles have formed on the edges of my habit and keep striking my legs, and blood flows from such wounds. And all covered with mud and cold, I come to the gate and after I have knocked and called for some time, a brother comes and asks: "Who are you?" I answer: "Brother Francis." And he says, "Go away; this is not a proper hour for going about; you may not come in." And when I insist, he answers: "Go away, you are a simple and a stupid person; we are so many and we have no need of you. You are certainly not coming to us at this hour." And I stand again at the door and say, "For the love of God, take me in tonight." And he answers: "I will not. Go to the Crosier's place and ask there." I tell you this: if I had patience and did not become upset, there would be true joy in this, and true virtue and the salvation of the soul. FRANCIS OF ASSISI

Receive poverty, want, sickness, and all miseries joyfully from the hand of God, and with equal joy receive consolation, refreshment, and all super-abundance. By this uniform joy in the will of God thou wilt deaden the stimulus of thy passions.   MACARIUS THE ELDER

Anyone who has the light to understand what I mean and the grace to follow it will experience, indeed, the delight of the Lord's playfulness. For like a father frolicking with his son, he will hug and kiss one who comes to him with a child's heart.   ANONYMOUS, *The Cloud of Unknowing*

Oh marvelous, indescribable, and joyful love, in Thee is all savour and sweetness and all delight, the contemplation whereof exalteth the soul above the world, making it able to stand alone in joy, peace, and rest.
          BL. ANGELA OF FOLIGNO

Be merry, really merry. The life of a true Christian should be a perpetual jubilee, a prelude to the festivals of eternity.   BL. THEOPHANE VENARD

Leave sadness to those in the world. We who work for God should be lighthearted.   LEONARD OF PORT MAURICE

Happiness is the natural life of man.   THOMAS AQUINAS

Since happiness is nothing other than the enjoyment of the highest good and since the highest good is above, no one can be happy unless he rise above himself, not by an ascent of the body, but of the heart.
          BONAVENTURE

So, abandon yourself utterly for the love of God, and in this way you will become truly happy.   BL. HENRY SUSO

The faithful person lives constantly with God, he is always serious and joyous: serious because he remembers God, joyous because he dreams of all the good things that god has given to man.   CLEMENT OF ALEXANDRIA

If the world knew our happiness, it would, out of sheer envy, invade our retreats, and the times of the Fathers of the Desert would return when the solitudes were more populous than the cities.   MADELINE SOPHIE BARAT

The heart is rich when it is content, and it is always content when it desires are fixed on God. *Nothing* can bring greater happiness than doing God's will for the love of God.   BL. MIGUEL FEBRES CORDERO-MUNOZ

So, brethren, "rejoice in the Lord," not in the world. That is, rejoice in the truth, not in the wickedness; rejoice in the hope of eternity, not in the fading flower of vanity. That is the way to rejoice. Wherever you are on earth, however long you remain on earth, "the Lord is near, do not be anxious about anything."   ANTHONY OF EGYPT

It is always springtime in the heart that loves God.   JOHN VIANNEY

My life is one of unending joy.   GABRIEL POSSENTI

The soul of one who loves God always swims in joy, always keeps holiday, and is always in a mood for singing.   JOHN OF THE CROSS

Laugh and grow strong.   IGNATIUS OF LOYOLA

I will have no sadness in my house.   PHILIP NERI

I want no long-faced saints.   JOHN BOSCO

A sad saint is a sorry saint.   FRANCIS DE SALES

Just as Joshua and Caleb held both that the Promises Land was good and beautiful and that its possession would be sweet and agreeable, so too the Holy Spirit by the mouths of all the saints and our Lord by his own mouth assures us that a devout life is a life that is sweet, happy, and lovable.
                                                          FRANCIS DE SALES

Melancholy is the poison of devotion. When one is in tribulation, it is necessary to be more happy and more joyful because one is nearer to God.
                                                           CLARE OF ASSISI

The monks have no sadness. They wage war on the devil as though they were performing a dance.   JOHN CHRYSOSTOM

I do not believe it would be possible to find any joy comparable to that of a soul in purgatory, except the joy of the blessed in paradise—a joy which goes on increasing day by day, as God more and more flows in upon the soul, which He does abundantly in proportion as every hindrance to His entrance is consumed away.   CATHERINE OF GENOA

If all divine consolations, all spiritual joys, all heavenly delights which ever were in this world—if all the saints who have ever lived from the beginning of the world until now were to expound and show forth God, if all the worldly delights, both good and evil, which ever existed were all to be converted into one good and spiritual joy which should endure until I were made perfect, I would not even that I might obtain all this, give or exchange even for the space of the twinkling of an eye that joy which I have in the unspeakable manifestation of God.   BL. ANGELA OF FOLIGNO

DEATH

Death is part of the curse of Adam, consequence of the Fall. "Dust thou art, and unto dust shalt thou return" (Genesis 3:19). Death was not designed to be natural and inevitable, but is rather the hideous fruit of sin. "I have set before thee this day life and good, and death and evil" (Deuteronomy 30:15).

The Incarnation changed the meaning of death forever. In Christ, immortal God had tasted death and in so doing destroyed death forever, "that through death he might destroy him who had the power of death, that is, the devil" (Hebrews 2:14) "for since by man came death, by man came also the resurrection of the dead. For as in Adam all die, even so in Christ shall all men be made alive" (1 Corinthians 15:21–22).

The saints thus can face bodily death with tranquility, and even with ardent longing, knowing that Christ has conquered death and what waits for them is a closer union with the object of their desire. "O death, where is thy sting? O Grave, where is thy victory?" (1 Corinthians 16:55).

God would be untrue if man did not die once God had pronounced sentence of death. On the other hand it would be unseemly that beings who were once made rational who have partaken in his Word should perish and return, by disintegrating, to non-existence. It were better not to have been created than, having come to exist, to perish through neglect. ATHANASIUS

What wholesome receipt this is. "Remember," saith this bill, "thy last things, and thou shalt never sin in this world." Here is first a short medicine containing only four herbs, common and well known, that is to wit, death doom, pain, and joy,   THOMAS MORE

Now in writing I confess it unto thee, O Lord! Read it who will, and interpret how he will; and if he finds me to have sinned in weeping for my mother during so small a part of an hour—that mother who was for a while dead in my eyes, and who had for many years wept for me, that I might live in thine eyes—let him not laugh at me, but rather if he be a man of noble charity, let him weep for my sins against thee, the Father of all the brethren of Thy Christ.   AUGUSTINE OF HIPPO

If it so please my Maker, it is time for me to return to him who created me and formed me out of nothing when I did not exist. I have lived a long time, and the righteous Judge has taken good care of me during my whole life. The time has come for my departure, and I long to die and be with Christ, my king, in all his glory.   BEDE THE VENERABLE

Life is given us that we may learn to die well, and we never think of it! To die well we must live well.   JOHN VIANNEY

Many a man in better health than you enjoy, more conservative in his personal habits, has gone to bed without a care and was not alive when morning came. Do not put your soul in such peril. By the love of Jesus Christ, by the Blood he shed to redeem your soul, make ready to give a good account of yourself and of all that God our Lord has given you to dispense.
                                    IGNATIUS OF LOYOLA

I shall be beheaded. Within a few short hours my soul will quit this earth, exile over, and battle won. I shall mount upwards and enter into our true home. There among God's elect I shall gaze upon what eye of man cannot imagine, hear undreampt of harmonies, enjoy a happiness the heart cannot comprehend. But first of all, the grain of wheat must be ground, the bunch of grapes trodden in the wine press.   BL. THEOPHANE VENARD

The Lord himself will take care of this lowly body of mine as befits his providence, whether this means unending suffering or some small consolation. Why am I anxious? "The Lord is near." But my hope is in his compassion that he will not delay in putting an end to this course which he has assigned to me.   MARTIN I, POPE

The approach of death is indeed the best news I could hear. A man must once pay the forfeit of death, and I do not value this life at a farthing if only our Lord will give me a tiny corner in paradise; nor do I worry about the future of the order, for I trust God to raise up men to assist and defend it.   CAMILLUS DE LELLIS

To order one's life properly, one must always think of being able to end it with a peaceful conscience, for it is more fearful and terrible to fall into the hand of the eternal Judge than to avoid the most intense bodily sufferings.   NICHOLAS OF FLÜE

As your soul departs from your body, may the shining cohorts of angels hasten to greet you, the tribunal of apostles acquit you, the triumphant ranks of white-robed martyrs accompany you, the lily-bearing bands of glorious confessors surround you, the choir of virgins bring up your train with rejoicing, and in blest tranquility may the patriarchs receive you into their loving embrace. May our Lord Jesus appear before you gentle and eager of countenance and assign you a place amid those who stand in his presence for evermore.   PETER DAMIAN, *to a dying friend*

Oh bride and fair one, oh thou who art beloved of Me with perfect love, of a truth I would not that thou shouldst come unto Me with these exceeding great sufferings, but I wouldst thou shouldst come with the utmost rejoicing and with joy unspeakable.   BL. ANGELA OF FOLIGNO

The whole of this present world and all that belongs to it—yes, this mortal life itself—has become mean and wearisome, and on the other hand the world to come, that shall not pass, and that eternal life have become so unspeakably desirable and dear that I hold all these passing things as light as thistledown. I am tired of "living."   HERMAN OF REICHENAU

Jesus, destroy this chain of a body, for I shall never be content till my soul can fly to you. When shall I be completely blessed in you?
                                        GEMMA GALGANI

Bridegroom and Lord, the longed-for hour has come! It is time for us to see one another, my Beloved, my Master. It is time for me to set out. Let us go.   TERESA OF AVILA

Much sorrow for the dead is either the child of self-love or rash judgement.   ROBERT SOUTHWELL

If God hears my prayers there will be no last words of mine to repeat, for I shall say nothing at all.   MADELEINE SOPHIE BARAT

To almost all the questions that might be asked about you the answer would be "perhaps." Shall you have a large fortune, great talents, a long life? "Perhaps." Will your last hour find you in the friendship of God? "Perhaps." After this retreat, shall you live long in a state of grace? "Perhaps." Shall you be saved? "Perhaps." But shall you die? "Yes, certainly."
                                        IGNATIUS OF LOYOLA

Often fill your mind with thoughts of the great gentleness and mercy with which God our Savior welcomes souls at death, if they have spent their lives in trusting Him, and striven to serve and love Him. Do your utmost to arouse in yourself a love of heaven and the life of the blessed so you will weaken your dread of parting from this mortal and fleeting life.
                                        FRANCIS DE SALES

I am waiting patiently for the day when I complete my offering to God. It will mean actually giving my life, but I don't mind leaving this world. My thoughts are above and beyond it all. I feel that my sojourn here is just about at an end. I am getting nearer and nearer to our real homeland. Earth is slipping away and Heaven is drawing near   BL. THEOPHANE VENARD

Do you not know that life on earth is a life of death? It is a living death and a dying life; a life which should be called death and not life; and earthy life an imperfect life, a sinful life. Will you say then that I am dead because I have left that miserable existence to participate in a heavenly life, a perfect life, an eternal and happy life? Do you not know that only the thoughtless and insane consider the faithful departed to be dead?   JEAN EUDES

After my death I shall let fall a rain of roses.   THÉRÈSE DE LISIEUX

HEAVEN

Heaven is the place and condition of perfect supernatural happiness, where the saints and angels enjoy the immediate vision and love of God. "Now are we the sons of God, and it doth not yet appear what we shall be, but we know that, when he shall appear, we shall be like him; for we shall see him as he is" (1 John 3:2). Although the same God will be seen and loved by all, each soul's capacity for participation will depend on merit.

Until the last judgement and final resurrection, only the souls of the just are in Heaven. After the last day their incorruptible, glorified bodies will rise also and be united with their souls. "And when the chief shepherd shall appear, ye shall receive a crown of glory that fadeth not away" (1 Peter 5:4).

To human minds, an important part of the promise of heaven is community; happiness is shared with angels and saints and with those who have been known and loved on earth. The blessed will experience the divine wisdom directly. What is only hinted at here will be fully comprehended. "For now we see through a glass, darkly; but then face to face: now I know in part; but then shall I know even as I am known" (1 Corinthians 13:12).

I have been made for Heaven and Heaven for me.　　　JOSEPH CAFASSO

Our home is—Heaven. On earth we are like travellers staying at a hotel. When one is away, one is always thinking of going home.

JOHN VIANNEY

The access to Heaven is through desire. He who longs to be there really is there in spirit. The path to Heaven is measured by desire and not by miles.

ANONYMOUS, *The Cloud of Unknowing*

Resolve henceforth to keep Heaven before your mind, to be ready to forego everything that can hinder you or cause you to stray on your journey there.

FRANCIS DE SALES

O Diana, what a wretched state of affairs this is, which we have to endure! Our love for each other here is never free from pain and anxiety. You are upset and hurt because you are not permitted to see me the whole time, and I am upset because your presence is so rarely granted me. I wish we could be brought into the city of the Lord of Hosts where we shall no longer be stranded from him or from each other.

BL. JORDAN OF SAXONY, *to Bl. Diana of Andalo*

Those four innocent souls that had made their way to Heaven before me, surely they would be sorry for their sister, sorely tried on earth. They could prove that love doesn't end in death. Before long a delicious sense of peace flooded into my soul and I realized that there were people who loved me in Heaven too. Ever since this devotion of mine to my little brothers and sisters has grown stronger; I have often sought out their company to tell them what a sad thing exile is, and how I long to join them soon in my true country.

THÉRÈSE DE LISIEUX

The love of all spirits is measured: and for this reason their love perpetually begins anew, so that God may be loved according to His demand and to the spirits' own desires. And this is why all blessed spirits perpetually gather themselves together and form a burning flame of love, that they may fulfill this work, and that God may be loved according to His nobility.

BL. JAN VAN RUYSBROECK

I see Paradise has no gate, but whosoever will may enter therein, for God is all mercy and stands with open arms to admit us to His glory. But still I see that the Being of God is so pure (far more than one can imagine) that should a soul see in itself even the least mote of imperfection, it would rather cast itself into a thousand hells than go with that spot into the presence of the Divine Majesty.

CATHERINE OF GENOA

Will He not give us all things when we are with Him? What shall our life and our nature not be when His promise unto us shall have been fulfilled! What will the spirit of man be like when it is placed above every vice that masters and subdues—when, its warfare ended, it is wholly at peace?

AUGUSTINE OF HIPPO

It is not said, "May the joy of thy Lord enter into thee," but "Enter thou into the joy of thy Lord," which is a proof that the joy will be greater than we can conceive. We shall enter into a great sea of divine and eternal joy, which will fill us within and without, and surround us on all sides.

ROBERT BELLARMINE

Heaven is not divided by the number of those who reign, nor lessened by being shared, nor disturbed by its multitude, nor disordered by its inequality of ranks, nor changed by motion, nor measured by time.

BONAVENTURE

I confess that I am bewildered and lose myself at the thought of the divine goodness, a sea without shore and fathomless, of God who calls me to an eternal rest after such short and tiny labors—summons and calls me to Heaven, to that supreme Good that I sought so negligently, and promises me the fruit of those tears that I sowed so sparingly.   ALOYSIUS GONZAGA

When we rise again with glorious bodies, in the power of the Lord, these bodies will be white and resplendent as the snow, more brilliant than the sun, more transparent than crystal, and each one will have a special mark of honor and glory, according to the support and endurance of torments and sufferings, willingly and freely borne to the honor of God.

BL. JAN VAN RUYSBROECK

How great, how lovely, how certain is the knowledge of all things there, with no error and no trouble, where the wisdom of God shall be imbibed at its very source with no difficulty and with utmost happiness!

AUGUSTINE OF HIPPO

MYSTIC UNION

"Mystic Union" is a foretaste of what the blessed souls in Heaven are experiencing. It is the union of the soul with God in deep contemplation, marked by a profound awareness of the divine presence. It is a supernatural state, wrought by the action of God, and no amount of human striving can produce it. There is a temporary transportation to the supernatural life as the result of an outpouring of unmerited grace so that human and divine meet in a union of love, "wherein he hath abounded toward us in all wisdom and prudence; having made known unto us the mystery of his will, according to his good pleasure which he hath purposed in himself" (Ephesians 1:8–9).

My beloved ones are encompassed by my love, and are absorbed into the one only thing; a love without images and without spoken words. They are gathered into me, the good from which they flowed.   BL. HENRY SUSO

Thérèse had simply disappeared, like a drop lost in the ocean. Jesus only was left, my Master, my King. Hadn't I begged him to take away my liberty because I was so afraid of the use I might make of it; hadn't I longed, weak and helpless as I was, to be united, once for all, with that divine Strength?   THÉRÈSE DE LISIEUX

The air is in the sunshine and the sunshine in the air. So likewise is God in the being of the soul; and whenever the soul's highest powers are turned inward with active love, they are united with God without means, in a simple knowledge of all truth, and in an essential feeling and tasting of all good.   BL. JAN VAN RUYSBROECK

I am his daughter, He said so. Oh, infinite gentleness of my God! Oh, word so long desired, so urgently besought! Ocean of joy! "My daughter!"
MARGARET OF CORTONA

My Lord and my God, take from me all that separates me from thee! My Lord and My God, give me everything that will bring me closer to thee! My Lord and my God, protect me from myself, and grant that I may belong entirely to thee!   NICHOLAS OF FLÜE

If you desire to know how these things come about, ask grace, not instruction; desire, not understanding; the groaning of prayer, not diligent reading; the Spouse, not the teacher; God, not man; darkness, not clarity; not light, but the fire that totally inflames and carries us into God by ecstatic unctions and burning affections.   BONAVENTURE

By virtue of love is the lover transformed in the beloved and the beloved is transformed in the lover, and the like unto hard iron which so assumeth the color, heat, virtue, and form of the fire that it almost turneth into fire, so doth the soul, united with God through the perfect grace of divine love, itself almost become divine and transformed in God.

BL. ANGELA OF FOLIGNO

To be loved by God, to be united to God, to live in the Presence of God, to live for God! Oh! how wonderful life is—and death!   JOHN VIANNEY

I implore Him to annihilate you entirely and to establish Himself perfectly in you; to draw and consume you completely within Himself; to be all in you, that one may no longer see anything but Jesus in your interior or exterior life, in time and eternity; to dwell in you, live and function in you, suffer and die in you, adore and glorify Himself in you in whatever way He desires.   JEAN EUDES

Let us die to ourselves; let us impose silence in solicitude, on desire, on the phantom of the senses. And in the train of the Crucified, let us pass from the world to our Father.   BONAVENTURE

Die! Die as the silkworm does when it has fulfilled the office of its creation, and you will see God and be immersed in His Greatness, as the little silkworm is enveloped in its cocoon. Understand that when I say "you will see God," I mean in the manner described, in which He manifests Himself in this kind of union.   TERESA OF AVILA

In order to become united with Him, a man must empty and strip himself of every inordinate affection of desire and pleasure for all that can be distinctly enjoyed, whether it be high or low, temporal or spiritual, to the end that the soul may purged and clean from all desires, joys, and pleasures whatsoever, and may thus be wholly occupied, with all its affections, with loving God.   JOHN OF THE CROSS

Then souls, arising with anxious desire, run, with virtue, by the bridge of the doctrine of Christ crucified and arrive at the gate, lifting up their minds in Me, and by My blood and burning with the fire of love they taste in Me, the eternal Deity, Who am to them a sea pacific with whom the souls have made so great union, and they have no movement except in Me. And being yet mortal, they taste the good of the immortals, and having yet the weight of the body, they receive the joy of the spirit. CATHERINE OF SIENA

Exalted, drawn, and absorbed into the uncreated Light, I beheld that which cannot be related.   BL. ANGELA OF FOLIGNO

The soul does not reflect on her own intelligence and will. She is completely buried in the abyss of the divinity. There she holds herself in silence, she sleeps, she rests in ineffable sweetness.   BL. HENRY SUSO

Then the soul is in God and God in the soul, just as the fish is in the sea and the sea in the fish.   CATHERINE OF SIENA

BIOGRAPHICAL NOTES

BL. AGOSTINA PIETRANTONI (1864–1894). Member of a nursing order of nuns, Sr. Agostina was murdered by a violent patient in a tuberculosis ward after a life of service nursing the poor. She forgave her killer as she died and was acclaimed a martyr.

ALBAN ROE (1583–1642). Born in Bury St. Edmunds, converted at Cambridge, and trained at Douai, Belgium, Fr. Roe returned to England as a missionary, and was hanged, drawn, and quartered for being a Catholic Priest. He is one of the Forty Martyrs of England and Wales canonized by Paul VI in 1970.

ALBERT THE GREAT (1206–1280). Son of a German count, Albert entered the new mendicant order of Dominican friars against his family's wishes and, with Thomas Aquinas, his pupil, became the order's most brilliant scholar. Albert's gifts and interests were wide ranging, encompassing physics, astronomy, chemistry, botany, geography, geology, physiology, logic, metaphysics, and ethics. He was one of the first and among the greatest of natural scientists and is counted as a Doctor of the Church.

ALEXANDER BRIANT (d. 1581). Another of the Forty Martyrs of England and Wales, Briant was born in Somerset and converted while a student at Oxford. He came to London from Douai in 1581 to proselytize, was captured, tortured for a year, and executed at Tyburn with fellow Saints Edmund Campion and Ralph Sherwin.

ALOYSIUS GONZAGA (1568–1591). Born to a powerful Italian family during a period of greater than ordinary violence and licentiousness, young Aloysius turned his back on the spirit of the times with vehemence to practice extreme austerity and sacrifice as a Jesuit. He died while ministering to plague victims and received the last rites from St. Robert Bellarmine, who later testified to his holiness.

ALPHONSUS LIGUORI (1696–1787). Trained as a lawyer, Alphonsus left the bar to become a priest after losing an important case through his own oversight. He organized the Congregation of the Most Holy Redeemer (Redemptorists) to conduct missions to rural districts. A theologian, a very popular writer on devotional subjects, and an ecstatic, Liguori's life was nonetheless filled with external struggles: theological combat against the puritanical Jansenist heresy and political conflicts over control of the Redemptorists. He is a Doctor of the Church.

AMBROSE (340–397). Son of a praetorian prefect of Gaul, Ambrose was appointed governor of the Milan district by the Emperor Valentinian. Ambrose, not yet a baptized Christian, went to Milan cathedral to settle a dispute between Arians and Catholics and found himself unanimously elected bishop by the warring factions. Ambrose went on to become a fierce defender of the independence of the Church against secular authority and a great builder of Christianity within the dying Roman Empire. He baptized his successor, St. Augustine. Ambrose is one of the four great Latin Fathers of the church, with Saints Augustine, Jerome, and Gregory the Great.

AMMONAS THE HERMIT (c. 288–350). One of the early desert fathers associated with Saint Anthony of Egypt, Ammonas established a monastery in the wilderness of Nitria, near Alexandria.

VEN. ANDREW BELTRAMI (1870–1897). One of the last recruits to the Salesian order clothed by St. John Bosco himself, young Beltrami made an early and fierce resolution to strive for holiness and perfection, which he carried out consistently during a short life filled with sickness and suffering.

ANDREW OF CRETE (660–740). Monk, archbishop, forceful preacher, liturgical poet, and hymnist, Andrew is the author of the *Great Kanon* penitential hymn for Lent still sung in the Byzantine liturgy.

BL. ANGELA OF FOLIGNO (1248–1309). One of the great Christian mystics, Angela lived a frivolous and worldly life until middle age, when a vision prompted her to give away all her possessions and to live a hermit's life of penance, austerity, and devotion to the poor. Her extraordinary visionary experiences were recorded and published by her confessor, a Franciscan friar.

ANGELA MERICI (c. 1470–1540). Born in Brescia, Lombardy, and orphaned as a child, Angela was a fearless and independent woman, a staunch traveler and pilgrim, a tireless teacher, and founder of the Ursulines, the first great teaching order for women.

BL. ANNA MARIA TAIGI (1769–1837). Wife of a domestic servant and mother of seven children, Anna Maria deeply impressed her confessor, Father Angelo, a Servite, with her life of heroic virtue and her spiritual gifts of vision and prophesy. His testimony led ultimately to her beatification, in 1920.

BL. ANNE JAVOUHEY (1798–1851). Anne's special mission was the education of poor children in Africa, South America, Tahiti, Madagascar, and Europe. She founded the Congregation of St. Joseph of Cluny.

ANSELM OF CANTERBURY (1033–1109). Born in Aosta, Italy, Anselm made his way to the abbey of Bec in Normandy, where the famous Lanfranc was abbot. The English clergy elected him to succeed Lanfranc as archbishop of Canterbury in 1092, in which position Anselm launched a heroic chapter in the perennial struggle to keep the English throne from dominating the Church. A great theologian, Anselm founded the Scholastic movement, which incorporated Aristotelian dialectics into theology, and proposed the most famous proof of the existence of God.

BL. ANTHONY GRASSI (1592–1671). A famous confessor, noted for his ability to read consciences and to see into the future, Grassi was Superior of the Oratorians at Fermo, Italy.

ANTHONY MARY CLARET (1807–1870). A weaver in early life, Claret became a tireless missionary and reformer of the church, founder of the Missionary Sons of the Immaculate Heart of Mary, archbishop of Santiago, Cuba, and chaplain to the Queen of Spain, whom he followed into exile after the Spanish revolution of 1868.

ANTHONY OF EGYPT (251–356). Born to prosperous Christian parents near Memphis, Egypt, Anthony gave away his inheritance and withdrew to the desert to a life of prayer, penance, and the legendary struggles with the devil described for us by St. Athanasius and, from a different angle, by Flaubert. The "founder of monasticism" had many imitators and tremendous fame and prestige throughout the civilized world. Saints and emperors journeyed to his mountain cave to seek his counsel until his death at the age of 105.

ANTHONY OF PADUA (1195–1231). A Portugese aristocrat, Anthony left his Augustinian convent in Lisbon after meeting a band of ragged Franciscans headed for martyrs' deaths in North Africa. Anthony became a Franciscan and did get to Morocco, but ill health forced him to return to Italy. He became the most famous preacher of his day, called, for his fearlessness and eloquence, "the hammer of the heretics." Charming stories of his miracles abound, like the famous episode of preaching to the entranced fishes when the hard-hearted populace refused to listen.

ANTHONY ZACCARIA (1502–1539). Anthony began life as a physician but gave medicine up to become a priest and found the Barnabites, an order dedicated to reform and to the revival of spirtuality in the Church.

ATHANASIUS (c. 297–373). Bishop of Alexandria and leader in the bitter struggle against the Arian heresy, Athanasius suffered repudiation, banishment, and attempted assassination. He was a friend and warm supporter of the work of St. Anthony of Egypt, whose biography he wrote in addition to a number of important treatises which earned him the title of "the champion of orthodoxy." He is a Doctor of the Church.

AUGUSTINE OF HIPPO (354–430). A black Numidean, raised as a Christian by his mother, St. Monica, but early converted to Manicheanism, Augustine suffered tormenting conflicts between the lure of the world and the call to holiness. He finally accepted baptism in 387 at the hands of St. Ambrose and went on to become the most brilliant exponent of the Christian view of history. A literary genius, and a most able bishop and administrator, he molded the structure of the church in Africa and the mind of the church in the West. He is the greatest of the Latin Fathers of the Church.

BASIL THE GREAT (329–379). Basil was born in Cappadocia in Asia Minor to a very good family; his grandmother, sister, two brothers, mother, and father are all numbered among the saints. He studied in Athens where St. Gregory Nazianzen and (for variety) the future emperor, Julian the Apostate, were classmates. Basil was the founder of Eastern communal monastic life and later became a most vigorous bishop of Caesarea and a leader of the anti-Arian faction.

BEDE THE VENERABLE (673–735). Raised from the age of three in the monastery of St. Peter and St. Paul at Wearmouth, Jarrow, Bede spent his life as a holy scholar. He is the father of English historians and, to a great extent, of English literature, and was the first to use *Anno Domini* to date events.

BENEDICT JOSEPH LABRE (1748–1783). After failing in a number of attempts to join a regular religious order, Benedict spent his life in perpetual pilgrimage, as a humble, ragged, filthy mendicant, wandering across Europe from Compostella to Loreto. After 1774 he settled down in Rome, sleeping in the Colosseum ruins and spending his days in churches. He is a striking example of the wandering holy man, a type of "fool for Christ's sake," more familiar to Eastern than to Western Christianity.

BENILDUS (1805–1862). A model of the selfless, gifted teacher, Pierre Romançon spent most of his life teaching little boys at the Christian Brother's school at Sauges, France. Pope Pius XI called him "the saint of the daily grind."

BERNADETTE SOUBIROUS (1844–1879). Born in poverty, humble, simple, and in poor health, Bernadette was chosen to experience the great visions and revelations of the Virgin at Lourdes, in the French Pyrenees, now the greatest modern center of religious pilgraimages from all over the world. Bernadette was canonized not for her visions, but for the patience and trust with which she bore the trials resulting from them.

BERNARD OF CLAIRVAUX (1090–1153). The commanding figure of his age, with immense influence over kings, popes, and the course of history, Bernard sought a purified, reformed, more austere monastic life and began by expanding the tiny Cistercian order in Clairvaux, Burgundy. His genius also expressed itself in his brilliant theological writing, which greatly influenced Christian mysticism and devotion to the Virgin. Bernard roused all Europe to the Second Crusade, led the attacks against Abelard's brand of rationalism, helped to stop pogroms, and established the legitimacy of Pope Innocent III against the antipope. Bernard is considered to be the last of the Fathers of the Church.

BERNARDINO OF SIENA (1380–1444). Great preacher and reformer of the Franciscan order.

BONAVENTURE (1221–1274). Most renowned theologian of the Franciscan order, Bonaventure was supposedly given his name by St. Francis himself, who cured him of a childhood illness. Bonaventure became General of the Friars Minor, Cardinal Archbishop of Albano, and one of the greatest medieval philosophers. He is a Doctor of the Church.

BONIFACE (c. 680–754). Born in Devonshire, and christened "Winfrid," Boniface was sent by Pope Gregory II to evangelize the pagan Germans. He was a great success; he converted a huge gathering at the Oak of Thor in Geismar and went on to found many monasteries and to be made primate of Germany.

BRIDGET OF SWEDEN (1303–1373). A Swedish noblewoman, mystic and visionary, foundress of the Bridgettine nuns. Bridget lived for many years in Rome, offering outspoken advice to popes, princes, and dignitaries of all degrees.

BRUNO (c. 1030–1101). Founder of the pure, austere, and sequestered Carthusian order ("never reformed because never deformed") in the desolate Alpine setting of *La Grande Chartreuse*.

BRUNO SERUNKUMA OF UGANDA (d. 1886). One of the Twenty-Two Martyrs of Uganda, St. Bruno was a bodyguard of the dissolute and tyrannical King Mwanga. He chose to die with the condemned royal pages who were to be executed for professing Christianity and for refusing to participate in the royal vices.

CAESARIUS OF ARLES (470–543). Born in Burgundy of a Gallo-Roman family, Caesarius entered the famous monastary of Lérins at the age of eighteen. His uncle Eonus, bishop of Arles, had the boy transferred to his diocese and made him his successor. As bishop, Caesarius fought the Arian and Pelagian heresies, and instituted the practice of having the Divine Office sung daily in all the churches of Arles. He was the first Western European bishop to receive the pallium, from Pope St. Symmachus. Caesarius published an adaptation of Roman law which became the civil code of Gaul.

CAJETAN (1480–1547). A creative and energetic reformer of the Church during an age when reform was much needed, Cajetan was co-founder of the Theatines and included the establishment of benevolent pawnshops for the poor among his innovations.

CAMILLUS DE LELLIS (1559–1614). Camillus spent his early years as a soldier of fortune and an addictive gambler. Illness and suffering opened his eyes to the needs of the sick, and he founded a congregation of priests and lay brothers for nursing work in hospitals, battlefields, homes, and hospices.

CATHERINE OF GENOA (1447–1510). Catherine and her husband were converted from a worldly life in Genoese society and devoted themselves to nursing the sick poor. Catherine's mystical experiences were frequent and intense. Her teachings received in visions were recorded by her confessor and have had considerable influence on religious thought.

CATHERINE LABOURÉ (1806–1870). A farmer's daughter from Burgundy, Catherine was a modest, unimaginative, inconspicuous Sister of Charity who received a dazzling series of visions which led to the immensely popular cult of the Miraculous Medal. Catherine herself lived and died in holy obscurity.

CATHERINE DEI RICCI (1522–1590). A Dominican nun and noted visionary, Catherine received the stigmata and a ring from Christ as the sign of her mystical marriage, which appeared to others as a red circle on her finger. She was a supporter of Savonarola, her fellow Florentine.

CATHERINE OF SIENA (1347–1380). The youngest of the twenty-five children of a Sienese dyer, Catherine began to have mystical experiences at the age of six. She resisted all of her family's efforts to persuade her to make the good marriage that her beauty and high spirits made a natural goal, devoting herself instead to prayer and good works. She attracted wide attention and a band of followers. Her reputation for holiness grew enormously, and she became very influential, successful urging, among other things, that Gregory XI leave Avignon and return to Rome.

CHARLES BORROMEO (1538–1584). Charles' appointment as cardinal at age twenty-two by his uncle, Pope Pius IV, proved to be one of the most successful acts of nepotism in history; Charles was a model of all the virtues, as well as a brilliant administrator, innovator, and reformer in an age when the corrupt Church needed such help desperately.

VEN. CHARLES DE FOUCAULD (1858–1916). The life of the Vicomte de Foucauld would be a fit subject for a Verdi opera. He spent the first half of his life as a rich, cynical rake, always on the point of being thrown out of his fashionable school or regiment for his many mistresses and blatant dissipation. A tour in the Moroccan desert stirred something within Charles, and the silently loving example of a cousin completed his conversion. He felt called to an apostolate of "presence, not preaching" and returned to the wilderness to live as a poor hermit, "little brother Charles of Jesus," among the fierce desert tribes.

BL. CHRISTOPHER BUXTON (d. 1588). Born in Derbyshire, Buxton was ordained in Rome, sent on the English mission, and hanged, drawn, and quartered at Canterbury for being a priest.

CHROMATIUS OF AQUILEIA (d. c. 407). Chromatius was a patron and friend of St. Jerome, a supporter of John Chrysostom, a scholar and author of several scriptural commentaries.

CLARE OF ASSISI (1194–1253). A daughter of the local nobility, the eighteen-year-old Clare was so inspired by hearing St. Francis preach that she ran away from home to join him. Her mother, two sisters, and other notable ladies eventually joined the group, which became the Poor Clares. Clare obtained from Pope Innocent III a special privilege guaranteeing the Poor Clares' right to hold no property at all. Second only to that of Francis, Clare's influence is responsible for the remarkable flourishing of the Franciscan movement.

CLEMENT OF ALEXANDRIA (c. 150–c. 217). Titus Flavius Clemens was of Greek origin and was probably raised as a pagan. He succeeded St. Panthenus as head of the Christian school at Alexandria, where Origen was one of his pupils. He was a prolific and widely read writer, whose theology was much influenced by Gnosticism and by the desire to reconcile as well as to contrast Christian and Gnostic ideas.

CLEMENT OF ROME (d. c. 99). Clement succeeded Cletus to become the third successor of St. Peter as bishop of Rome. He was exiled and persecuted by the Emperor Trajan and, according to the legend, eventually lashed to an anchor and thrown into the sea. A well-authenticated letter from Clement to the Corinthians provides the earliest known example of the Roman bishop intervening with authority in the affairs of another church.

COLUMBA (c. 521–597). Born in Donegal, of royal descent, Columba went off to evangelize the heathen Picts of Scotland. He and twelve companions founded the monastery on the island of Iona which became the greatest in Christendom, a center of learning and a main source for the spread of St. Benedict's rule.

COLUMBANUS (540–615). The most influential of all the monks from Ireland, Columbanus spent his life founding abbeys and monastic centers all over Western Europe, from Fontaines to Bobbio. He ran into considerable opposition to his insistence on the stern Irish rule, which was eventually, indeed, replaced by the more gentle Benedictine code.

CUTHBERT OF LINDISFARNE (d. 687). An ardent missionary, who roamed throughout northern England, Cuthbert in his later years withdrew to his beloved island of Lindisfarne to live as a hermit until unwillingly made a bishop. He spent the last part of his life visiting his flock and performing prodigies of healing and prophecy.

CYPRIAN OF CARTHAGE (c. 200–258). Cyprian was a pagan Carthaginian intellectual who, once converted, became a profound biblical scholar and bishop of Carthage. He was beheaded during the Valerian persecutions for refusing to sacrifice to the old gods.

CYRIL OF ALEXANDRIA (c. 376–444). Cyril's life as a theologian and leader of the church is marked by a contentiousness and rigor that seem the reverse of saintly meekness. His chief service to the Church was his vigorous defense of what were to become the orthodox doctrines of the Trinity and the Incarnation against the Nestorian and Pelagian heresies.

CYRIL OF JERUSALEM (c. 315–386). As bishop of Jerusalem, Cyril was deeply embroiled in fighting the Arian heresy, which proposed that the Son is less than the Father. He was repeatedly expelled from his see during various phases of the struggle but remained deeply attached to Jerusalem and its preservation.

DAVID LEWIS (1616–1679). Raised as a Protestant in Wales, Lewis became a Catholic in Paris and entered the English College in Rome, returning to Wales in 1648 to a farmhouse at Cwm that functioned for thirty-one years as the center for Jesuit activities in the west of England. He was captured during the Titus Oates plot persecutions and was hanged, drawn, and quartered. He is one of the Forty Martyrs of England and Wales.

DOMINIC (1170–1221). A Castilian, Dominic spent his early years living a studious and contemplative life as an Augustinian Canon at Osma, until his bishop chose him as a companion on a mission to Languedoc, where the

Albigensian or Catharist heresy was firmly established. Dominic saw the need to combat heresy by better teaching of Christian truths. He decided to establish an order of dedicated, knowledgeable preachers who could explain the gospels accurately, simply, and forcefully and who would set an example of Christian virtue in their style of living. Before his death, the Order of Friars Preachers was flourishing all over Europe.

DOMINIC SAVIO (1842–1857). A peasant boy, Dominic joined St. John Bosco in Turin at the age of twelve and died three years later, at fifteen. Don Bosco, who was deeply impressed with Dominic's goodness and his spiritual gifts of discernment, compassion, and prophecy, wrote the boy's biography.

EDMUND CAMPION (1540–1581). Raised as a Catholic in London, Campion took the Oath of Supremacy, acknowledging Elizabeth I as head of the Church in England, and accepted preferment from her, then later recanted, fled to Douai, and joined the Jesuits. He was one of the first sent on the English mission. Amidst the uproar and excitement aroused by the publication of his apologia ("Campion's Brag"), he was captured and hanged, drawn, and quartered. He is one of the Forty Martyrs of England and Wales.

EDMUND THE MARTYR (841–870). Chosen as King of the East Angles at the age of fourteen, Edmund ruled wisely until defeated and captured by the invading Danes at Hoxen, Suffolk. When he refused to share his Christian kingdom with the pagan invaders, he was tied to a tree, shot full of arrows until he "looked like a thistle," and then beheaded.

ELIZABETH OF HUNGARY (1207–1231). Daughter of the King of Hungary, and wife of a German prince, Ludwig II, Landgrave of Thuringia, Elizabeth was turned out of the Wartburg castle by her brother-in law when her husband was killed on a Crusade. She renounced the world, joined the third order of Saint Francis, and became legendary for her loving care of the poor and needy.

ELIZABETH SETON (1774–1821). The first American-born saint, Elizabeth Bayley came from a well-connected New York family. Left a widow with five children, she was ostracized by her family and friends when she became a Catholic in 1805. She moved to Baltimore to found a religious community, the Sisters of St. Joseph, and lay the foundations for the parochial school system in the United States.

EPHRAEM OF NISIBIS (c. 306–373). Ephraem's many writings include poetry, theology, and well-known hymns. The only Syrian Doctor of the Church, he was a great champion of the Virgin and of the Immaculate Conception and a scourge of the Arians and the Gnostics.

EUSEBIUS OF VERCELLI (c. 283–371). Bishop of Vercelli and originator of the custom of a bishop living with his canons in community, Eusebius was banished and persecuted by Emperor Constantius for his firm stand in support of Athanasius as opposed to Arius. He was one of the authors of the Athanasian Creed.

FELICITY OF CARTHAGE (d. 203). The eyewitness account, possibly written by Tertullian, of the martyrdom of Felicity and Perpetua and four male companions, is one of the most complete, touching, and well-authenticated of all the narratives of early Christians being thrown to the wild beasts.

FIDELIS OF SIGMARINGEN (1577–1622). Born in Prussia, Fidelis practiced law in Alsace, then gave his goods away and became a Capuchin monk. He was successfully preaching and evangelizing the Swiss Protestants when he was assassinated.

FRANCES OF ROME (1384–1440). Frances Ponziana was a Roman aristocrat who throughout her long and happy marriage was also drawn to a life of charity and self-denial. She founded a community of Roman ladies who were dedicated to doing good works while living "in the world." They nursed plague victims and sold their jewels to help the poor. Frances was guided during the last decades of her life by an archangel, visible only to her, and performed many miracles of healing and prophecy.

FRANCES XAVIER CABRINI (1850–1917). The youngest of thirteen children of a Lombard farmer, Frances was trained as a schoolteacher, then tried to enter a religious order and was twice rejected. With the encouragement of her bishop, she founded the Missionary Sisters of the Sacred Heart and was invited to New York in 1889 to work with Italian immigrants. She went on to found more than fifty hospitals, schools, and orphanages all over the United States, Central and South America, and in Italy and England. Frances was the first American citizen to be canonized. She is the patroness of emigrants.

FRANCIS OF ASSISI (1181–1226). The most universally known and ecumenically beloved of all the saints, Francis was born to a wealthy silk merchant and spent his youth in pursuit of pleasure and of glory in war, of which Francis had an idealized, chivalric image. Captured in a battle between Assisi and Perugia, he was imprisoned and later suffered a serious illness which changed his temperament. He experienced compelling visions of Christ and began to devote his life to serving the sick and the poor. After a dramatic public break with his father, who now considered him mad, Francis turned to a life of absolute poverty and simplicity, wandering through the world calling all to the practice of charity and penance. In an astonishingly short time, he attracted many saintly followers and the "friars minor" were spread throughout the known world. Francis himself went to Egypt to evangelize the Mohammedans, and to Palestine. In 1224, while praying on Mount Alverna in the Apennines, he received the stigmata, the five wounds of the crucified Christ. After Francis's death the movement he had founded became divided, inevitably, into purist and more compromising and conventional branches.

FRANCIS BORGIA (1510–1572). A redeeming member of the infamous Borgia family, Francis was a model duke (of Gandia, in Spain), incorruptible and pious. After the death of his wife, he became a Jesuit and turned his estates over to his eldest son. He was an enormous success as a preacher and an energetic founder of houses throughout Spain and Portugal. In 1561 he was made Governor General of the Jesuits and was so effective he was called their "second founder" after St. Ignatius.

BL. FRANCIS LIBERMANN (1804–1852). The son of an Alsatian rabbi, Francis was first drawn to the Church by reading Rousseau's arguments for the divinity of Christ in *Emile*. He joined the Sulpicians to train for the priesthood in Paris, but his ordination was delayed for a number of years when he developed epileptic seizures. He faced this and numerous other emotional and spiritual trials with a transcendent faith. He was drawn to the goal of serving the black peoples of the world, particularly in Africa, and founded a congregation that merged with an older, nearly defunct order to become "The Holy Ghost Fathers." Libermann understood the needs of mission superbly well and persuaded the French and the pope to agree to appoint African bishops to allow for all-important adaptations to local needs.

FRANCIS OF PAOLA (c. 1416–1507). Founder of the "Friars Minim" in an effort to return to the absolute poverty and simplicity characteristic of the original Franciscan spirit, Francis slept on a rock and asked his followers to observe a perpetual Lenten fast. He was called from Calabria to France by Louis XI, then kept there by Louis's heir Charles VIII, who would do nothing without the advice of the holy man. Charles built several monasteries and the Roman church of Santa Trinità del Monte for the friars of Francis's order.

FRANCIS DE SALES (1567–1622). One of the most eloquent and gifted saints, Francis was born in the family chateau at Thorens, Savoy, and forsook the possibility of a brilliant secular career to enter the Church. He was first sent to bring back the people of his own region, who had become Calvinists, to the Church. He succeeded by preaching with love and with the spiritual insight that also characterized his writings. He served as a wise and charitable Bishop of Geneva. St. Jeanne Françoise de Chantal founded the Order of the Visitation with his guidance and encouragement. St. Francis is a Doctor of the Church and the patron of journalists and writers.

FRANCIS XAVIER (1506–1552). Francis was born, like St. Ignatius of Loyola, to the Basque nobility. The two men met at the University of Paris, and Francis became one of the first seven Jesuits in 1534. He subsequently traveled as a missionary to Portugese India, to Japan, and to China, where he died, after many hardships heroically endured and many miracles of conversation and healing.

FULGENTIUS OF RUSPE (468–533). Bishop of Ruspe in what is now Tunisia, Fulgentius was exiled by Thrasimund, King of the Vandals, who adhered to the Arian heresy. Fulgentius spent half the years of his episcopacy aboard. He wrote a number of treatises against Arianism and in defense of orthodoxy.

GABRIEL POSSENTI (1836–1862). One of the saints who were canonized because of their exemplary purity and humility, St. Gabriel was a Passionist monk who died at the age of twenty-four. He is a patron of Youth.

GEMMA GALGANI (1878–1903). Gemma was a stunning Tuscan girl of humble background who bore the stigmata and experienced extraordinary visions, diabolical assaults, and ecstasies, which prompted both persecution and the development of a popular cult which eventually led to her canonization.

GERARD MAJELLA (1726–1755). Son of a tailor, Gerard became a lay brother in St. Alphonsus Liguori's Redemptorist Congregation. Simple and uneducated, Gerard had the gift of humility and a dazzling array of spiritual gifts, including bilocation (being seen in two places simultaneously), healing, reading of souls, ecstacy, and prophecy.

GERTRUDE THE GREAT (c. 1256–1302). Raised by St. Gertrude of Hackeborn, St. Mechtilde, and the nuns of the convent of Helfta in Saxony, Gertrude became a scholar and a mystic. She produced many books inspired by the revelations she received in visions and ecstasies.

BL. GILES OF ASSISI (d. 1262). One of St. Francis's first followers, Giles is a dominant figure in the collection of tales called *Fioretti* or *The Little Flowers of St. Francis*. He traveled widely, from Compostella to Tunisia to Palestine, and was Francis's companion on many missions in Italy.

GREGORY I THE GREAT (c. 540–604). One of the greatest of the popes, and the father of the medieval papacy, Gregory called himself "the servant of the servants of God." It was Gregory who, according to Bede, saw the blond, barbarian English for sale as slaves in Rome, pronounced them "not Angles but Angels," and sent St. Augustine of Canterbury to evangelize them. He established the independence of the papacy from the Byzantine emperor and vastly increased the spiritual and temporal powers of the Pope.

GREGORY II (d. 731). A very active pope, Gregory rebuilt the walls of Rome to ward off the Saracens, fought heresies, reformed clerical discipline, and sent St. Boniface and St. Corbinian to Germany to convert the pagans. His diplomacy helped keep the Lombard threat in check, and he fought the Iconoclast movement with vigor.

GREGORY III (d. 741). A Roman priest with a great reputation for holiness, Gregory was acclaimed pope while accompanying the funeral procession of his predessor, Gregory II. He turned the papacy toward the West and laid the foundations of the Holy Roman Empire when he sought the aid of Charles Martel against the Lombards, rather than that of the emperor in Constantinople.

GREGORY NAZIANZEN (c. 329–389). Gregory's father, mother, and two brothers are also saints. He, St. Basil the Great, and St. Gregory of Nyssa are called "the Cappadocian Fathers" and were chiefly responsible for the final defeat of the Arian heresy. Gregory is a Doctor of the Church and is also called "the Theologian" for his writings, particularly on the Trinity.

GREGORY OF NYSSA (c. 330–395). Brother of St. Basil the Great, who suggested that he be appointed bishop of Nyssa, this Gregory spent his career fighting the Arians and was given the labels of "Pillar of Orthodoxy" and "Father of Fathers" by church councils.

HEDWIG OF SILESIA (c. 1174–1243). Daughter of the Count of Bavaria, Hedwig married the heir to the Silesian throne, and used her position to sponsor the building of churches, hospitals, monasteries, to promote peace in eastern Europe, and to help the poor. She is the aunt of St. Elizabeth of Hungary, who had a similar career.

BL. HENRY SUSO (c. 1295–1365). A Dominican monk, son of a German nobleman, Suso was a mystic and a visionary theologian. He called himself "Servant of the Eternal Wisdom." He had a difficult life and suffered much from malicious accusations. He was a renowned spiritual director and preacher.

BL. HERMAN OF REICHENAU (1013–1034). Called "The Cripple" because of his severe physical deformities, Herman was placed in the Abbey of Reichenau in Lake Constance at the age of seven, and spent his whole life there. He was a mathemetician, a poet, a historian, and author of the hymn *Salve Regina*, among others.

HILARION (c. 291–371). Born in Gaza and educated in Alexandria, Hilarion spent time in the desert with St. Anthony and developed a taste for the life of the hermit. He spent subsequent years fleeing from one spot to another, trying to escape the consequences of his own celebrity. The solitude which he sought always eluded him because of his reputation for spectacular holiness and miracle-working.

HILDEGARD (1098–1179). Called "the Sybil of the Rhine," Hildegard was a visionary and prophetess, consulted by popes, emperors, kings, and other saints. She was prioress of a Benedictine convent near Bingen.

HUGH OF LINCOLN (1140–1200). Born at Avalon in Burgundy, Hugh entered the austere Carthusian order founded by St. Bruno and was called to England by Henry II to become abbot of the Carthusian monastery founded by the King as part of his penance for having ordered the death of St. Thomas Becket. Hugh later made an admirable bishop of the See of Lincoln and was a notable reformer, protector of the Jews, and upholder of the rights of the Church against the demands of the Crown.

IGNATIUS OF ANTIOCH (d. c. 107). According to legend, Ignatius was the little child shown by Christ to his apostles. He was later appointed Bishop of Antioch by St. Peter. He was arrested during the persecutions of the Emperor Trajan and sent to Rome to be killed by the wild beasts, probably in the Colosseum.

IGNATIUS OF LOYOLA (1491–1556). Born to the Basque nobility in northeastern Spain, Ignatius, like Francis of Assisi, spent his youth soldiering, "given over to the vanities of the world," and was ardently in love with chivalry until wounded in battle and left with time to think during convalescence. He had an intense conversion experience, becoming a mystic with the desire to enlighten the world. His mysticism still bore the stamp of the soldierly discipline of his youth, as did the Jesuit order which he founded. The Jesuits were dedicated to reforming the church (especially through education), to missionary activities, and to the fight against heresy.

ILDEPHONSUS (607–667). Archbishop of Toledo, and Marian enthusiast, Ildephonsus wrote an early treatise on the perpetual virginity of Mary.

IRENAEUS (c. 125–c. 203). The first great Catholic theologian, one generation removed from the apostles, Irenaeus was close to St. Polycarp, who had been a pupil of St. John. Polycarp sent Irenaeus to Gaul, where he was an active missionary and served as Bishop of Lyons. He wrote a famous treatise against the Gnostic heresy then prevalent in Gaul.

ISAAC JOGUES (1607–1646). One of the heroic Jesuit missionaries who were canonized as the Martyrs of North America in 1930, Jogues worked successfully with the Hurons in Quebec, was captured by the Iroquois, and imprisoned and tortured for a year, and was finally beheaded by the Mohawks, at Auriesville, New York.

ISIDORE OF SEVILLE (c. 560–636). Bishop of Seville, encyclopedist, and Doctor of the Church, Isidore was a prolific writer. His works on history, religion, and science were very popular during the Middle Ages.

JEAN BAPTISTE DE LA SALLE (1651–1719). Head of a rich and noble family of Reims, La Salle was living a leisured, dignified, and scholarly life as a canon of the cathedral when he began helping Adrien Nyel, who was struggling to open a school for poor boys. He became intensely involved in the project, resigned his fortune and canonry, and devoted the rest of his life to founding and guiding the Institute of the Christian Brothers, a congregation of lay brothers devoted to the education of poor students, which had a tremendous impact on the early development of modern education.

JEAN DE BREBEUF (1593–1649). Born in Normandy, a farmer's son, Brebeuf joined the Jesuits and was one of the first three missionaries sent to the Huron Indians in Canada, among whom he labored for twenty-four years. He made seven thousand conversions and wrote a dictionary and catechism in the Huron language. He was captured by the Iroquois, bitter enemies of the Hurons, cruelly tortured, and killed.

JEAN EUDES (1601–1680). Another sturdy and tough-minded son of a Norman father, Eudes joined the Congregation of the Oratory and became a leading missionary and opponent of the puritanical Jansenist heresy. He founded a refuge for fallen women, run by the Visitandines, then went on to found his own congregation (Eudists) of secular priests dedicated to improving seminaries and to preaching missions.

BL. JAN VAN RUYSBROECK (1293–1381). Born near Brussels and educated by his uncle, who was a canon, John, his uncle, and another canon lived as contemplatives and hermits, eventually forming a community of Canons Regular of St. Augustine after the number of their disciplines grew larger. Ruysbroeck's mystical writings are of the first rank and have been enormously influential.

JEANNE FRANÇOISE DE CHANTAL (1572–1641). Grandmother of Mme de Sévigné, the Baroness of Chantal was an unhappy widow in her thirties when she came under the influence of St. Francis de Sales. At his urging she founded a new religious order (the Order of the Visitation) for single women and widows who were not fit for the severe life of other orders but who wanted to live as religious.

JEROME (c. 342–420). The greatest scholar of the early Church, and a fiery controversialist, Jerome produced the translation of the Bible called the Vulgate, which remained the official Latin text of the Bible until 1979. Jerome was baptized by Pope Liberius, learned Hebrew at Antioch from a rabbi, and studied in Constantinople with St. Gregory Nazianzen. He finally settled in Bethlehem, with St. Paula, St. Eustochium, and other disciples, where they founded monasteries, schools and a hospice, worked and studied, and welcomed refugees from troubled areas.

JEROME EMILIANI (1481–1537). One of a significant number of soldier saints, Jerome was commander of the forces in a town that fell to the Venetians. After capture and imprisonment, he later became a priest. He established hospitals, a home for repentant prostitutes, and a congregation devoted to caring for widows and orphans.

BL. JOHN OF ALVERNA (1259–1322). John joined the Franciscans at age thirteen and became a hermit at Monte Alverna. He was a famous preacher and spiritual adviser throughout central and northern Italy.

JOHN OF AVILA (1499–1569). Spiritual adviser to St. Teresa of Avila, St. John of the Cross, St. Francis Borgia, and St. Peter of Alcantara, this John followed the Scriptures without compromise, giving away his fortune to the poor and suffering imprisonment by the Inquisition for his attacks on the powerful and corrupt.

JOHN BOSCO (1815–1888). One of the most lovable of the saints, warm-hearted and very human, John Bosco spent his life working to reclaim, house, educate, and encourage boys from the slums of Turin and later from backward areas all over the world. He often had to rely on miracles to feed his unruly hordes, and luckily was generally able to produce them when required. There are well attested stories of his multiplying food and of money turning up just as hundreds of his boys were about to be evicted. He wrote biographies of two saints, one of his mentor, Joseph Cafasso, and one of St. Dominic Savio, one of his pupils who died at fifteen. He called his foundation Salesian, after St. Francis de Sales, whom he much admired.

JOHN CHRYSOSTOM (c. 347–407). An eloquent preacher (chrysostom means "golden-mouthed"), John was made patriarch of Constantinople against his wishes, in which role he made many enemies among the powerful by denouncing their ostentation and lack of charity. A synod of thirty-six more deferential bishops recommended exile, and John spent his last three years writing hundreds of letters from his outpost in Armenia. He is a Doctor of the Church and the patron of preachers.

JOHN CLIMACUS (c. 509–649). John was a monk and abbot at Mount Sinai, and wrote a guide to attaining perfection, *The Ladder of Paradise*, which was very popular throughout the Middle Ages.

JOHN OF THE CROSS (1542–1591). Son of a Castilian silk weaver, John became a Carmelite monk but felt called to a life of more intense prayer and contemplation. He told St. Teresa of Avila of his wish to join the Carthusians, but she persuaded him instead to help her with her plans to reform the Carmelites. The reformers met with strong resistance, and John was imprisoned and mistreated, first by the unreformed and later by the extremists of the reforming party. He wrote some of his most beautiful poems in prison. John of the Cross as a poet, mystic, and spiritual psychologist is known and read even by many who have no interest in Christianity as such. He is a Doctor of the Church.

JOHN DAMASCENE (c. 675–c. 749). Called "the last of the Eastern Fathers," John was the son of a Christian official at the caliph's court in Damascus and spent all his life under Muslim rule. He eventually inherited his father's post and became counselor to Abdul Malek. He became a monk and was deeply involved in the controversy with the Byzantine emperors over their efforts to prohibit the veneration of images (iconoclasm). His writings, especially *The Orthodox Faith*, were very influential, and for the elegance of his prose he was known as "chrysorrhoas" or "gold-pouring."

JOHN FISHER (1469–1535). Son of a Yorkshire textile merchant, Fisher became Chancellor of Cambridge University and chaplin and administrator for Lady Margaret Beaufort, mother of Henry VIII, then bishop of Rochester. From 1527 he consistently upheld the validity of Henry's marriage to Catherine of Aragon and opposed the king's claims to be head of the Church in England. While he was imprisoned in the Tower of London, Pope Paul II named him a cardinal, infuriating the king, who promptly had Fisher executed for treason.

BL. JOHN GABRIEL PERBOYRE (1802–1840). Born at Puech, France, Perboyre became a member of the Congregation of the Mission, founded by St. Vincent de Paul. He was tortured to death in China after being sold by one of his own converts for thirty pieces of silver.

JOHN OF GOD (1495–1550). Another soldier-saint, John also had worked as a slave overseer in Morocco, and as a shepherd in Seville. He became so violent a penitent after hearing a sermon by St. John of Avila that he was placed in a lunatic asylum. He emerged to help prostitutes, vagabonds, the poor, and the sick. John founded the Order of Brothers Hospitalers and is the patron of nurses, hospitals, and the sick.

JOHN VIANNEY (1786–1859). Born in the hamlet of Dardilly, near Lyons, to a devout peasant family, John lived through the suppression of the Church during the French Revolution. He was early attracted to the priesthood and studied with the local curé, but his own academic slowness, as well as political upheavals, greatly hampered him. When nearly thirty, he was at last ordained because his bishop was impressed by his devotion and simplicity, and because Lyons, like all of France, was desperate for priests after the ravages of the past twenty-five years. Vianney was assigned to the remote and dismal parish of Ars, where he served for the rest of his life and whose name he made famous. He possessed a remarkable gift for reading souls, and Ars became a place of pilgrimage for people from all over Europe and the Americas who besieged him in his confessional for up to eighteen hours a day. Vianney sought the serenity of a Carthusian monastery several times but was always discovered and dragged back to his parish to continue his ministry. He is the patron saint of parish priests.

BL. JORDAN OF SAXONY (d. 1237). Born in Germany and educated at Paris, Jordan succeeded St. Dominic as second Master of the Order of Preachers in 1222. Of a passionate, loving, and fervent nature, Jordan led the order into a period of expansion, founding and establishing many Dominican convents. He was shipwrecked on his way back from a pilgrimage to the Holy Land in 1237.

JOSEPH CAFASSO (1811–1860). Joseph was born in Castelnuovo d'Asti near Turin (now called "Castello Don Bosco" after another saintly villager) to devout peasant parents. His vocation to the priesthood and to a life of exemplary sanctity came early and never wavered. Called "the Priest's Priest," Cafasso, like St. John Vianney was famed as a confessor of supernatural discernment and irresistible persuasive powers. His techniques for dealing with Scruples led victims from all over Europe to seek him out. He had spectacular success converting the felons in Turin's prisons and felt his "santi impicatti," or holy hanged ones, were perhaps the parishioners nearest to his heart.

JOSEPH OF CALASANZA (1556–1648). A precursor of Jean Baptiste de la Salle in the realm of education of poor children, Calasanza was born in Spain, studied at three universities there, and went to Rome as assistant to Cardinal Colonna. He was shocked at the degradation of the Roman poor and established a number of free schools in the slums, with priests as teachers. This became the order of Clerks Regular of the Religious Schools or "Piarists." As so often in the careers of the great founders, his order was torn by dissention. St. Joseph was accused of maladministration, humiliated, supplanted, and deposed. He bore his trials with patience, and the order was reformed and restored after his death.

VEN. JOSEPH DE VEUSTER ("Father Damien") (1840–1889). De Veuster went as a missionary to Hawaii in 1864 and saw the pitiful state of the lepers in their colony at Molokai. From 1873 onward he lived among them as their spiritual and physical minister, not only praying for the sick and the dead but often building their coffins as well. He contracted the disease himself after years of living as a virtual prisoner among his flock. Father Damien suffered the usual trials of calumny and slander, but found a notable defender in Robert Louis Stevenson, who wrote a stirring account of his life.

JOSEPH OF COPERTINO (1603–1663). A poor boy of frail health and feeble wits, St. Joseph became a stable boy at a Franciscan monastery and improved so much in this setting that he was eventually taken as a Franciscan novice and ordained. He became famous, and even notorious, for the variety and the exuberance of his ecstasies, miracles, and supernatural gifts. His levitations were particularly striking, and there are over seventy eyewitness accounts by reliable and even royal observers of St. Joseph's flights. During his last years he was shunted about from one Franciscan house to another in an effort to impose obscurity on his doings and to discourage pilgrims. He is the patron of pilots and air travelers.

BL. JULIAN OF NORWICH (c. 1342–1423). One of the most eloquent and beloved of the English mystics, Julian became an anchorite outside the walls of St. Julian's church in Norwich after experiencing a series of revelations in 1373. By the time of her death, her reputation for sanctity attracted pilgrims from all over Europe to her cell.

JULIE BILLIART (1751–1816). Daughter of a well-to-do northern French farmer, Julie was highly religious and devoted to the sick and poor from her youth. She was paralyzed by a mysterious illness for twenty-two years, then cured instantly by a command from her spiritual adviser. She founded the Institute of Notre Dame for the spiritual education of poor children and the training of religious teachers. As is usual in the case of saintly founders, she was accused of misgovernment, was deposed as head of the order, and endured many trials and struggles before it could be reestablished.

JUSTIN MARTYR (c. 100–c. 165). Born in Palestine to a pagan Greek family, Justin studied philosophy and became a Christian, after being a Platonist, in his early thirties. He became a vigorous and able Christian apologist and traveled widely, expounding the gospel and debating with the pagan philosophers on the reconciliation of faith and reason. After being denounced as a Christian by a Cynic whom he had beaten in debate, he was tried by the Roman prefect and beheaded.

LAWRENCE GIUSTINIANI (1381–1455). A Venetian aristocrat who turned to a life of penance and austerity, Lawrence used to go about the city begging alms for the poor with a sack on his shoulder. He was unwillingly made Bishop and later Patriarch of Venice and was revered for his prophetic powers, his charity, and his success as a reformer of the church.

LAWRENCE O'TOOLE (1128–1180). Son of an Irish chieftain, Lawrence was captured in a raid and eventually turned over to the Bishop of Glendalough, where he became a monk, an abbot, and then archbishop of Ireland. During his tenure the first English forces invaded Ireland under the Earl of Pembroke, and for the rest of his life Lawrence was involved in mostly thankless efforts to reconcile disputes between Henry II of England, the Irish chieftains, and the pope.

LAWRENCE OF BRINDISI (1559–1619). A Doctor of the Church, St. Lawrence was a learned and vigorous Franciscan who preached many missions, particularly to Lutherans and Jews. He led troops into battle against the Turks invading Hungary, armed only with a crucifix. He founded friaries all over Europe and wrote many sermons, scriptural commentaries, and works of controversy.

LEO THE GREAT (d. 461). Leo was made pope during a period when Rome was besieged by heresy, Huns, and Vandals. He fought off all with authoritative courage and strengthened the influence of the Roman Church in Spain, Gaul, and North Africa. He is a Doctor of the Church.

LEONARD OF PORT MAURICE (1676–1751). Son of a mariner, Leonard joined the Franciscans and preached missions all over Italy for forty-three years. He developed and popularized the devotion of the Stations of the Cross. For a time he was spiritual adviser to Clementina Sobieska, wife of the Old Pretender, whose son, Cardinal Henry of York, promoted Leonard's canonization.

LOUIS-MARIE GRIGNION DE MONTFORT (1673–1716). Born in Brittany of poor parents, Louis was educated by the Jesuits and became a popular missionary in his native province, preaching in an emotional and flamboyant style which aroused opposition as well as enthusiasm. He wrote a number of very popular devotional works devoted to the Virgin and to the Rosary and founded two religious congregations, the Daughters of Divine Wisdom and the Missionaries of the Company of Mary.

BL. LOUIS GUANELLA (1842–1915). The ninth of thirteen children in a hardworking family of Campodolcino, in the Italian Alps, Louis was influenced by the work of John Bosco and the Salesian tradition of practical charity. His Servants of Charity founded homes for the retarded (whom Guanella called his "good children"), reclaimed swamplands for charity, rescued earthquake victims, and helped Italian immigrants to the United States.

LOUIS IX OF FRANCE (1214–1270). A saintly and effective king, "the Ideal Monarch," Louis led armies on crusade, dealt with the encroaching English both militarily and diplomatically, strengthened the French throne, and acted the role of peacemaker and protector of the weak in domestic and foreign politics. Louis built Sainte Chappelle in Paris to house Christ's Crown of Thorns, given to him by Emperor Baldwin II, and supported the founding of the Sorbonne.

BL. LYDWINA OF SCHIEDAM (1380–1433). A Dutch mystic and visionary, Lydwina suffered an injury while skating that left her a lifelong invalid. She dedicated her suffering to God and received many supernatural gifts. She is the patron saint of ice-skaters.

MACARIUS THE ELDER (c. 300–390). An Egyptian cattleherder, Macarius became a hermit and withdrew to the desert of Skete to escape public attention. He was exiled to a small island in the Nile by the Arian Patriarch of Jerusalem, but he eventually returned to the desert and lived there for sixty years.

MADELEINE SOPHIE BARAT (1779–1865). Daughter of a Burgundian cooper, Madeleine was given an excellent, if strict, education by her older brother, who was in holy orders. He recommended her to his friend Fr. Joseph Varin for the role of foundress of an order to educate girls on the model developed by the Jesuits. It became the enormously successful Society of the Sacred Heart of Jesus. Before her death, Madeleine had established more than one hundred houses and schools in twelve countries.

MARGARET OF CORTONA (1247–1297). Another saint whose life would make a fine operatic libretto, Margaret was a beautiful Tuscan peasant girl who ran away from an unsympathetic stepmother, caught the eye of a young nobleman, and lived luxuriously as his mistress until he was murdered in 1273. Margaret's shock at this event led to her conversion and to a public renunciation of her past sins. She and her young son were taken in by charitable women, given spiritual support by Franciscan friars, and under their direction Margaret became a spectacular penitent, ecstatic, visionary, and foundress of charities, much sought after for her spiritual counsel and her miraculous healing powers.

BL. MARIA CHIARA OF CHINA (1872–1900). One of Seven Protomartyrs of the Franciscan Missionaries of Mary in China who were killed in the massacres of Christians during the Boxer Rebellion. The seven nuns are part of a larger group of twenty-nine martyrs, beatified together by Pius XII in 1946, called Blessed Gregory and Companions.

BL. MARIA DROSTE ZU VISCHERING (1863–1899). One of seven children of a German Count, Maria had an early, strong vocation and became an exemplary nun and superior in the Order of the Good Shepherd. She received a revelation that the pope was to consecrate the world to the Sacred Heart, and her appeals to the pope were successful. She died, as she had predicted, a few hours after receiving the papal encyclical enacting the dedication.

MARIA GORETTI (1890–1902). Maria was a simple, hardworking, pious girl who lived with her poor, widowed mother in the Pontine marshes near Nettuno, Italy. She was brutally murdered by the son of her landlord while resisting his attempted rape. Maria lived long enough to declare her forgiveness and concern for her murderer, who was converted in prison and released in time to testify at the hearings for Maria's beatification in 1929 and to receive communion at her shrine. His case has been cited as an argument for the abolition of capital punishment.

MARTIN I (d. c. 656). The latest pope to be venerated as a martyr, Martin was born in Umbria and came to Rome to serve as nuncio to Pope Theodore I, whom he succeeded. He infuriated the Emperor Constans II by opposing the Monothelite heresy, which denied the humanity of Christ. Constans had Martin taken captive and brought to Constantinople, where he was imprisoned, then exiled to the Crimea, and there mistreated until he died.

BL. MARY AMADINE OF CHINA (1872–1900). Born in Herck-la-Ville, Belgium, Mary was hacked to death by the Boxers as one of the twenty-nine Companions of the Blessed Gregory in China.

BL. MARY FORTUNATA VITA (1827–1922). Born Anna Felice in Veroli, Italy, Bl. Mary cared for her dissolute father and many younger siblings after her mother's early death, then entered a Benedictine convent. Here she rejected a chance to receive an education and become a choir religious, preferring to practice humility as a lay sister and servant. She received a number of spiritual gifts, including that of prophecy, and served quietly till the age of ninety-five.

VEN. MARY MAGDALEN BENTIVOGLIO (1834–1905). Born in the Castel Saint Angelo, daughter of a count who was also a papal general, Annetta Bentivoglio was ordered by Pius IX to make a foundation of the Poor Clares in the United States. She died in Omaha, Nebraska after establishing two houses against heavy odds in a country which had little appreciation for the uses of the contemplative life.

MARY MAGDELENE DEI PAZZI (1566–1607). Another in a distinguished series of Carmelite saints and mystics, Catherine dei Pazzi came from a distinguised Florentine family. She had the gifts of healing, prophecy, and the ability to read souls. Records of her utterances while in ecstatic states were published by the sisters of her convent.

MAXIMILIAN KOLBE (1894–1941). A martyr saint to symbolize concerns appropriate to the second half of the twentieth century as Thérèse de Lisieux did for its first half, Kolbe was born in Poland. He became a Franciscan and founded the chivalrous Knights of Mary Immaculate to promote devotion to Mary. He established 'cities' and religious publications dedicated to Mary in Poland, India, and Nagasaki, Japan. He was arrested by the Gestapo after the German invasion of Poland and was sent to Auschwitz in 1941, where he volunteered to take the place of a married man with a family in a group of prisoners who were to be executed in retaliation for a prison escape. He is often portrayed behind barbed wire, clad in the stripes of a concentration camp prisoner.

BL. MICHAEL RUA (1837–1910). Successor to St. John Bosco as the "second father" of the Salesian order, Rua extended and expanded its activities, opening missions in eight new countries. He was particularly sensitive to the need to respect local culture and customs and urged his Salesians to enrich the practice of the Church by adapting and "sanctifying" them.

BL. MIGEL FEBRES CORDERO-MUNZO (1854–1910). Miguel was the first native Ecuadorean to become one of the de la Salle Christian Brothers. He was an exemplary teacher and scholar, prolific writer, and a model of humility and trust in God. Crippled as a young child, Miguel took his first steps at age five when he saw "a beautiful lady" beckoning to him.

NICHOLAS OF FLÜE (1417–1487). One of a scant number of Swiss saints, Nicholas was a staunch citizen, happily married patriarch, father of ten, magistrate, officer, and prosperous farmer. At the age of fifty he felt a call to the eremetical life and withdrew to a cottage near his old home where he lived for the last nineteen years of his life, eating nothing but the communion wafer. Known by the affectionate name of "Brother Klaus," he had visitors from all walks of life and was frequently consulted by the leaders of his own and other countries because of his reputation for wisdom. He proposed the terms for the Compromise of Stans in 1481 which prevented civil war in Switzerland.

NILUS THE ABBOT (c. 910–1004). Nilus, a Calabrian Greek, was turned from a worldly life at the age of thirty when the woman he lived with and their child died. He became a monk and was made abbot of San Demetrio Corone, with a great reputation for wisdom and holiness. Nilus was forced to flee to the north with many of his monks in 981 when the Saracens invaded southern Italy. They were given shelter at Monte Cassino. Nilus eventually went on to found several other houses.

NILUS SORSKY (c. 1433–1508). Of Russian peasant origin, Nilus lived for many years as a monk on Mount Athos in Greece studying monastic discipline and mysticism. He then established a colony of hermits in Russia devoted to translating and studying Greek mystical writings. Nilus led a movement of "nonpossessors" dedicated to austerity in monastic life and in church decoration.

PATRICK OF IRELAND (c. 385–471). Patrick was born to an upper-class Roman-British family but was kidnapped by Irish pirates at age sixteen and spent his formative years as a shepherd in the pagan wilds of western Ireland, where he developed an intense inner life of prayer and awareness of God. He escaped his captors and made his way to Gaul but returned to Ireland in 432 after being called in a dream to preach and evanglize there. He overcame the resistance of the Irish chieftains and Druids, organized the Irish church, and converted the island, making Ireland a beacon of the faith and of the young Christian civilization.

PAUL MIKI (1562–1597). One of the group canonized in 1862 as the Martyrs of Japan, Paul was the son of a Japanese military chief who was educated at the Jesuit college at Anziquiama, joined the Society in 1580, and became a famous orator and controversialist. He was crucified near Nagasaki along with twenty-five other Japanese Christians during the persecutions of Toyotomi Hideyoshi.

PAUL OF THE CROSS (1694–1775). Paul was born to impoverished noble parents in Ovada, Italy. Inspired by a vision, Paul and his younger brother John Baptist founded an order dedicated to preaching the Passion of Christ (The Passionists). Paul drew enormous crowds to his missions and converted many criminals and hardened sinners. He also had the gifts of bilocation, healing, and prophecy.

PAULINUS OF NOLA (353–431). Son of the Roman prefect of Gaul, Paulinus was taught poetry and rhetoric by the poet Ausonius. After the death of their only child, he and his wife devoted their lives to austerity and charity. He finally settled in Nola in southern Italy, where he lived as a hermit and cared for the shrine of St. Felix of Nola. The people of the district chose him for their bishop. Paulinus was a friend and correspondent of Saints Augustine, Jerome, Martin of Tours, and Ambrose. His distinguished letters and poetry have survived.

PEASANT OF ARS (19th century). An anonymous villager immortalized by the Curé of Ars, who repeated an anecdote about the old man's holy simplicity. He may stand for the saints who are nameless in this world.

PETER CANISIUS (1521–1597). Peter was the son of a burgomaster of Nijmegen, Netherlands. He was educated at Cologne as a lawyer but became inspired by the Jesuit, Peter Faber, and joined the Society in 1543. He became the most effective theologian, organizer, preacher, and writer of the Counter Reformation in southern Germany, Bohemia, and Austria. He has been called "the Second Apostle of Germany," St. Boniface being the first. He is a Doctor of the Church.

PETER CHRYSOLOGUS (406–c. 450). According to legend, Peter was named bishop of Ravenna by Pope St. Sixtus III in obedience to a vision, ousting another candidate who had been elected by the people. Peter reformed his see, stamped out paganism, and preached so eloquently that he was called Chrysologus or "the golden mouthed."

PETER CLAVER (1580–1654). A Jesuit from Catalonia, Peter was sent to South America and found his mission in ministering to the hideous sufferings of the African slaves who poured into the port of Cartagena. Claver worked in unspeakable conditions for thirty-three years, caring for their spiritual and physical needs. He died alone and neglected after four years of illness and pain. He is patron of all missionary activities to black peoples.

PETER DAMIAN (1001–1072). Peter was born in poverty at Ravenna, Italy, and spent his early youth as a swineherd. An elder brother who had become a priest rescued him and had him educated at the schools of Lombardy. He became a vigorous and outspoken reformer of the Church; one of his tracts against corrupt clergy was called "The Gomorrah Book." He was named a Doctor of the Church by Leo XII without previous formal canonization.

PETER JULIAN EYMARD (1811–1868). Born near Grenoble in the French Alps, he worked for some time as a parish priest, then joined the Marist fathers, becoming renowned as a confessor and preacher. He left the order in 1856 to found his own congregation, The Priests of the Blessed Sacrament, to foster devotion to the Eucharist. He also established the Servants of the Blessed Sacrament for nuns devoted to perpetual adoration.

PETER MARTYR (1205–1252). He was born in Verona, Italy of parents who had embraced the Catharist heresy. They gave him an orthodox education, however, and he became a Dominican at the time when that order was leading the fight against Catharism. Peter was appointed inquisitor of northern Italy and was very successful at fighting the heresy there. He was assassinated by a Cathar and later canonized as a martyr.

PHILIP NERI (1515–1595). A native Florentine who spent most of his life in Rome, Philip came from a middle-class family. He was educated by the Dominicans, rejected a business career after a brief trial, and developed a unique apostolate in Rome, preaching in the streets, making converts, attracting disciples from all walks of life, and renewing religious life in his chosen city. His personality was marked by a joyful eccentricity, love of laughter, gentleness, and humility. He gathered his followers into regular meetings to pray and hear sacred music, which became the Congregation of the Oratory.

BL. PHILIPPINE DUCHESNE (1769–1852). Philippine was born to a rich family in Grenoble, France, and joined the Visitation nuns. The community was dissolved during the Revolution, and she returned home and tried to rebuild a community after 1801 when the Church was restored in France. She persuaded St. Madeleine Sophie Barat to accept the convent for her Society of the Sacred Heart and went to the United States as a Sacred Heart Missionary to found free schools for Indians in Missouri and Kansas.

PIUS X (1835–1915). Born near Venice, to very poor parents, Joseph Sarto used to walk four miles to school barefoot to save shoe leather. He became a parish priest and rose through the ranks to cardinal. His pontificate was noteworthy for his advocacy of early and frequent communion, separation of the Church from political control, liturgical reform, and the condemnation of Modernism and of political extremism of the left and right. His life was marked by simplicity and poverty.

BL. PLACID RICCARDI (1844–1915). Placid was a Benedictine monk at the church of St. Paul's Outside the Walls at Rome. He was a much sought-after confessor, pastor to the poor, and chaplain to nuns, who led a quiet, exemplary life of service and obedience.

POLYCARP (c. 69–c. 155). Polycarp was converted to the faith by St. John the Evangelist and became Bishop of Smyrna. He and his friend, St. Ignatius of Antioch, form a link between the age of the apostles and the great Church fathers of the second century. He was burned at the stake in the region of Marcus Aurelius for refusing to sacrifice to the gods.

PROCLUS OF CONSTANTINOPLE (d. 447). A disciple of St. John Chrysostom, Proclus became patriarch of Constantinople in 434. He was notable for his gentle and considerate handling of Nestorians and other heretics and his dedication to the people of his see.

RAYMOND OF PEÑAFORT (1175–1275). A Spanish Dominican, Raymond became master general of his order and encouraged Thomas Aquinas to write his *Summa Contra Gentiles*. At the request of Pope Gregory IX, he codified the canon law in a form that remained authoritative from 1234 until 1917.

BL. RICHARD ROLLE (c. 1300–1349). Born at Thornton, Yorkshire, Rolle studied at Oxford and at the Sorbonne. He became a hermit, settling at Hampole where he was spiritual director of a community of Cistercian nuns. He wrote exceptionally beautiful and insightful accounts of the contemplative life in both Latin and English.

ROBERT BELLARMINE (1542–1621). Bellarmine was born in Montepulciano, Tuscany, and educated by the Jesuits. He entered the Society in 1560 and became a renowned scholar and teacher at the University of Louvain and at the Roman College. One of the ablest and staunchest defenders of the Church against Protestantism during the Counter Reformation, he was noted for the soundness and elegance of his reasoning rather than for rhetoric, emotionalism, or dogmatic assertion.

ROBERT SOUTHWELL (c. 1541–1595). Born at Horsham St. Faith's, Norfolk, Southwell was the son of a courtier. He was sent abroad to study at Douai and Paris, then became a Jesuit in 1578. He was sent on the English mission and worked as a priest in London from 1584 to 1592, when he was betrayed and captured. After three years of imprisonment and torture, he was hanged, drawn, and quartered at Tyburn. Southwell is the author of a number of well-known religious poems and prose works.

ROSE OF LIMA (d. 1586). The first saint of the New World and patroness of South America, Rose was born of Spanish parents in Peru. Seeking to model her life on that of St. Catherine of Siena, she became a Dominican tertiary, lived in a shack in the family garden, and practiced extraordinary penances. A learned commission appointed to investigate her mystical experiences and visions determined that they were of supernatural origin. She devoted much care to poor and sick Indians and slaves.

SEBASTIAN VALFRÉ (1629–1710). Born in Piedmont, Valfré joined St. Philip Neri's Congregation of the Oratory in Turin. He was famous as a pastor, spiritual director, and preacher of great insight and prophetic gifts.

SIMEON OF EMESA (d. c. 589). Also called Simeon the Insane ("Salus") because he deliberately assumed eccentric and outrageous behavior to court humiliation and to be considered a "fool for Christ's sake," Simeon lived as a hermit for thirty years in the Sinai Desert and in Syria. Among other things, he used to take goods from shops and give them to the poor.

SOPHRONIUS (d. c. 638). Born in Damascus, Sophronius, called "Sophist," or the wise, lived as a monk in Egypt, traveled around the Near East, was made patriarch of Jerusalem, and acted as the Calif Omar's guide to the sacred places when the Arabs took Jerusalem in 637.

TERESA OF AVILA (1515–1582). A lively, intelligent girl of good family, Teresa entered a Carmelite convent in Avila, Castile, as a young woman and found it a fairly easygoing place which allowed much contact with the outside world. Her own spirituality developed gradually through a remarkable series of mystical experiences well described in her writings, and she eventually became a reformer of the Carmelite order, establishing convents which were to adhere to strict enclosure, poverty, austerity, and a life of intense prayer. She faced opposition and criticism with courage, determination, and humor. Her books are spiritual classics, and she is a much-loved Doctor of the Church.

THEODORE OF HERACLEA (d. c. 306). Identified with possibly two separate early soldier-martyrs from Heraclea, near Pontus on the Black Sea, Theodore is esteemed as a martyr and as a warrior saint. In one version of his legend, he fought a dragon, like St. George. In another, he set fire to a pagan temple and was burnt in reprisal.

BL. THEOPHANE VENARD (1829–1861). A devout boy, Venard was born in Poitiers, France, and decided while very young to serve in the missions and to accept probable martyrdom. He was sent to the Far East in 1852, when he was twenty-three years old, and was beheaded at Ke Cho, West Tonkin (Vietnam) six years later. He wrote a touching series of letters to his family describing his experiences and his imprisonment.

THÉRÈSE DE LISIEUX (1873–1897). The most popular saint of the early twentieth century, Thérèse Martin, "the Little Flower," lived an obscure life, entering a Carmelite convent in Lisieux, Normandy at age fifteen, and dying there, of tuberculosis, at age twenty-four. Her older sister, who was prioress, ordered her to write an account of her life and spiritual development, which was published after her death and became a sensational success. In *The Story of a Soul*, Thérèse describes her practice of "the little way" of spiritual perfection. It is the way of humility, of small, ordinary deeds sanctified by being performed in a spirit of total love and trust in God. She is co-patroness of missions with Francis of Assisi, and of the French nation, with Joan of Arc.

THOMAS AQUINAS (c. 1225–1274). Born to a princely family in Sicily, Thomas was related to the Holy Roman emperor and the king of France. When he entered the newly formed Dominican order to become a mendicant friar, his horrified family kidnapped and imprisoned him in the family castle at Roccasecca for over a year. Released, he went to Cologne to study under the Dominican master St. Albert the Great and went on to teach at Paris, Naples, Anagni, Orvieto, Rome, and Viterbo. Aquinas became the greatest theologian of the Middle Ages. His work strives to reconcile faith and reason as far as possible, making clear both the limitations and the uses of reason. He was a gentle, prayerful man of great personal holiness, whose fame and genius never marred his humility. He is a Doctor of the Church.

THOMAS BECKET (1118–1170). A middle-class Londoner of Norman descent, Thomas studied law in Paris and joined the household of Theobald, Archbishop of Canterbury, who made him archdeacon of Canterbury and sent him on numerous important missions abroad. Thomas was a close friend of the king, Henry II, who made him chancellor, then, hoping to have a reliable man in the office, archbishop of Canterbury. This position affected Thomas's character and way of life; he became austere, devout, and zealous for the interests of the Church, which brought him into bitter conflict with the disappointed king. Thomas was murdered in the transept of his own cathedral by three of the king's knights who believed Henry had ordered them to get rid of the archbishop. The martyr's tomb at Canterbury became the second greatest medieval pilgrims' shrine in Europe, after that of St. James at Compostella in Northwest Spain.

THOMAS MORE (1478–1535). Another saint, like Anselm and Thomas Becket, who fell afoul of an English king in the perennial struggle between Church and state in Britain, More was trained as a lawyer and was Henry VIII's friend and chancellor. He was also a distinguished scholar, writer, and humanist, author of *Utopia*. More refused to support Henry's efforts to divorce Catherine of Aragon in order to marry Anne Boleyn and he declined to sign the Act of Supremacy declaring the king sole head of the Church in England. He was beheaded for treason, declaring that he died "the King's good servant, but God's first." More is the patron saint of lawyers.

THOMAS OF VILLANOVA (1488–1555). Archbishop of Valencia, noted for his generosity and care for the poor ("deserving and undeserving"), his spiritual gifts and powers of healing, and his eloquence in preaching.

BL. THOMAS WOODHOUSE (d. 1573). Ordained during the reign of Mary Tudor, when the Catholic Church was briefly restored in England, Thomnas became a parish priest but was forced to resign when Mary died. He was later arrested for saying Mass, was imprisoned for twelve years, and eventually executed for treason after asking Lord Burleigh to urge Elizabeth I to be reconciled with the pope.

VINCENT DE PAUL (c. 1580–1660). Vincent came from southwestern France of peasant stock. He lived a life filled with exciting incident and with boundless charity. He was captured by pirates and sold as a slave in Algeria. He escaped and, back in France, was sent on secret missions for King Henry IV and served as chaplain to the Queen. Vincent founded missions and foundations to help galley slaves, war refugees, the sick, poor peasants, decayed gentry, Protestants, Jansenists, and abandoned children. He worked for the ranson of Christian slaves in Africa and for the eradication of human suffering wherever he found it.

VINCENT FERRER (1350–1419). Born to an aristocratic family in Valencia, Spain, Vincent joined the Dominicans at seventeen and became a preacher of great power and persuasiveness. He was said to have the gift of tongues as he was able to move crowds with equal force whether or not he spoke the local language. A close friend of Peter de Luna, who became the second "antipope" in Avignon, Vincent later helped to end the Western Schism by withdrawing his allegiance from de Luna and persuading the king of Spain to do the same.

ZACHARIAS (d. 752). Of Greek background, Zacharias was born in southern Italy, became famous for his learning and holiness, and was elected bishop of Rome in 741. He dealt effectively with the invading Lombards through diplomatic means, promoted the cause of Pepin the Short as king of the Franks, and sponsored St. Boniface's mission to convert the Germans.